D1715659

INDOOR AIR QUALITY

SAMPLING METHODOLOGIES

INDOOR AIR QUALITY

SAMPLING METHODOLOGIES

Kathleen Hess-Kosa

Cover photo (Top) — bat guano component; insect body parts. (Bottom) — bat guano components; insect antenna. (Photos courtesy of Shawn Abbott, Ph.D., Laboratory Director, Environmental Microbiology Laboratory, Inc., Daly City, CA.)

Library of Congress Cataloging-in-Publication Data

Hess-Kosa, Kathleen.
Indoor air quality : sampling methodologies / Kathleen Hess-Kosa.
p. cm.
Includes bibliographical references and index.
ISBN 1-56670-539-8 (alk. paper)
1. Indoor air pollution—Measurement. I. Title.

TD890 .H49 2001
628.5′3—dc21 2001029590

Direct all inquiries to CRC Press LLC, 2000 N.W. Corporate Blvd., Boca Raton, Florida 33431.

Visit the CRC Press Web site at www.crcpress.com

Lewis Publishers is an imprint of CRC Press LLC

International Standard Book Number 1-56670-539-8
Library of Congress Card Number 2001029590
Printed in the United States of America 1 2 3 4 5 6 7 8 9 0
Printed on acid-free paper

PREFACE

This book is intended to provide environmental professionals and industrial hygienists with the latest information available on "indoor air quality sampling." The focus is to provide a "practical guide" for developing a theory and following it through to the identification of unknown air contaminants.

The Background Section shall define, clarify, and provide direction for developing a hypothesis. With a well-defined hypothesis, the investigator must then test the hypothesis by sampling. Direction is provided for determining what, when, where, and how to sample for the various airborne components that may be found in indoor air quality situations. The components are broken into bioaerosols, chemicals, and dust.

In each chapter after the background information, occurrence of materials, sampling strategies, and several methodologies have been discussed. The intent of providing a several sampling methods for a given substance is to allow the investigator to choose the appropriate approach for different scenarios. The objective is not that of being focused on the single best approach but to choose that which is most appropriate given differing situations. As with a detective, there is no single recipe to fit all cases.

Although generally defined as all biological components, bioaerosols are inclusive of microbials only in the Bioaerosols Section of this book. Other biological components are contained with the section on dust. This section contains sampling methodologies for microbial allergens, pathogenic microbes, and toxigenic molds/bacteria.

The Chemical Section contains sampling methodologies for volatile organic compounds, mold volatile organic compounds, carbon dioxide, carbon monoxide, formaldehyde, and product emissions. The mold volatile organic compounds are discussed within this section because some researchers speculate that the mold byproducts may contribute to the total volatile organic compound exposures in an enclosed building. Yet, others use some of the techniques provided in order to locate molds in wall spaces. Thus, when performing a mold assessment, the investigator should be aware of other approaches that may aid in diagnosing and identifying a source.

The Dust Section contains sampling methodologies for many of the allergenic bioaerosols as well as methods for identifying components of dust. Dust components can reveal not only allergenic materials, but chemicals adsorbed onto the surfaces of dust or settled droplets, toxic metals, and different types of fibers (e.g., fiberglass).

Overlooking the need to assess the dust component be overlooking very important information. It is not how much dust is indoor air but what is in it.

As a passion for detective work is one of the great motivators for one the focus on indoor air quality problems, the person performing an assessment is herein referred to as the "investigator." The investigators greatest asset is his/her ability to weave through a convoluted web of complex problems. This book provides a few strategic approaches and "tools."

Kathleen Hess-Kosa

ACKNOWLEDGMENTS

I wish to dedicate this book to those who have sacrificed time, technical expertise, and patience. Of particular note is David Fetveit who consistently bombarded me with the latest and greatest each time I thought the book was coming to a close. I would also like to thank David Gallup for providing me with the extensive information regarding last minute details. Shawn Abbot, Ph.D. responded to my never-ending questions and provided a wonderful set of micrographs of molds. Brian Sheldon, Ph.D. also assisted me in details that are difficult at best to obtain. Each of these individuals contributed toward the ever-changing, ever-challenging topic of bioaerosols.

Last, but certainly not least, Mike Kosa has given me support and suffered the loss of my presence during the final months of frantic deadlines and glaring retorts. He has been my rock of sanity.

ABOUT THE AUTHOR

Kathleen Hess-Kosa is president/owner of Omega Southwest Environmental Consulting. In 1972, she received her bachelor of science degree in microbiology with a minor in chemistry from Oklahoma State University. After serving as an officer in the Air Force for three years, she returned to school and earned a master of science degree (1979) in industrial hygiene from the College of Engineering at Texas A&M University. Her research involved an animal toxicological study and was conducted at the College of Veterinary Sciences.

Ms. Hess-Kosa worked as a consultant for Firemen's Fund Insurance Companies until 1984. This was shortly after she passed the certification exam conducted by the American Board of Industrial Hygiene. During these five years, Ms. Hess-Kosa had the opportunity to become involved in a variety of unique industrial hygiene and environmental concerns, including indoor air quality concerns in an 800-occupant office building and information gathering for performing environmental site assessments and assessing waste. All this and much more carried over to her private consulting business.

Ms. Hess-Kosa has since conducted numerous Phase I environmental site assessments and published a book concerning the topic. She has actively pursued obscure sources of information and training to better address the complex nature of environmental issues, indoor air quality, and multiple chemical sensitivity. She has successfully identified sources of indoor air quality problems in over 90 percent of the numerous investigations performed, and she has been instrumental in rectifying 100 percent of the scenarios. It took some time to get to this point, but some of the information which was collected and has been used by Ms. Hess-Kosa is presented within this book.

Table of Contents

Section I

BACKGROUND

Chapter 1

HISTORIC OVERVIEW

The Environmental Protection Agency (EPA) ranks indoor air pollution among the top four environmental risks in America. People spend about 90 percent of their lives indoors, and pollution is consistently 2 to 5 times higher indoors than outdoors. The indoor pollutant levels have been reported as high as 100 times the levels encountered outside.

Since the worldwide energy crisis in 1973, advances in energy efficiency building construction have not been without a down side. In an effort to conserve fuel in commercial and residential buildings, builders started constructing airtight buildings, inoperable-airtight windows, and reduced air exchange rates.

In well weatherized homes, the air exchange rate is presently 0.2 to 0.3 air changes per hour. In older, less energy efficient homes, exchange rates have been measured at 2 changes per hour. In energy efficient office buildings, air exchange rates are around 0.29 to 1.73 changes per hour. The higher exchange rates in older buildings dilute and clean indoor air contaminants, and the newer buildings retain them. Illness associated with new buildings has, thus, come to be referred to as tight building syndrome.

By 1986, the news media began to sensationalize the condition as sick building syndrome. Sick building syndrome infers a condition in which the occupants of a building experience acute health and comfort problems that seem to be linked to a building but whose cause is unknown.

Indoor air quality investigative methods have only recently become technologically sophisticated in regards to investigative approaches and analytical procedures. Initially, formaldehyde off-gassing from furnishings in office buildings and off-gassing from particleboard in mobile homes was targeted as the culprit. One article, published in 1987, refers to formaldehyde as a deadly sin. Others declared: "It could be your office that is sick," "Tight homes, bad air," and "The enemy within." All referenced formaldehyde.

With the passage of time, other substances were implicated (e.g., carbon monoxide, tobacco smoke, and organics), but all the environmental professionals really knew was that health complaints could generally be alleviated with greater air exchange rates. Thus, the poor air exchange in buildings has been implicated as the cause of many of the complaints for which there was no known source.

In the latter part of the 20th century, residential concerns were being addressed with greater frequency. The target components included carbon monoxide, allergens, electromagnetic radiation, radon, and a medley of household products.

Recently, the target-offending agents in all indoor air environments have become molds. Headlines herald, "The dish on hotel air," "Moldy attitudes on indoor air need a good scrubbing," "The good, the bad and the moldy," and "Fungal sleuths."

An estimated 1.34 million office buildings have problems with air quality, and approximately 30 percent of all office employees are potentially exposed to the health effects of poor indoor air quality.[1] Over 50 million Americans suffer from asthma, allergies, and hay fever. Chronic bronchitis and emphysema increased by greater than 85 percent between 1970 and 1987. Close to 100,000 Americans die each year because of complications due to chronic obstructive pulmonary diseases.

Over 50 percent of our nation's schools have poor ventilation and significant sources of pollution in buildings where an estimated 55 million students and school staff members are affected by poor air quality. Health effects are predominantly observed in children with asthma. In the last 15 years, a 60 percent increase in the incidence of asthma has occurred amongst school-aged children. Today, approximately 8 percent of all school-aged children have been diagnosed with asthma.

Many indoor air quality situations culminate with litigation, differences in health impact, differences in perceived health effects, and failed regulatory limits. Guidelines are being created by recognized public agencies, and investigators are being called upon to make decisions with minimal support and direction. Indoor air quality investigations are truly only now becoming common.

LITIGATION

Managing indoor air quality has become one of the more demanding challenges facing school administrators and potentially office facility managers. Legal action, negotiation, and arbitration have re-defined what is considered as acceptable. An acceptable response to indoor air quality complaints has, thus, come to be defined in terms of reasonable standard of care.

If a student or faculty member initiates an indoor air quality claim against a school, the person must establish certain facts. First, the claim must demonstrate that the school has a duty to protect faculty from reasonably foreseeable harm. Second, after having demonstrated the existence of a duty, the claimant must demonstrate that the school failed to provide a reasonable standard of care. This constitutes negligence. For example, if a staff member reports health symptoms associated with the building and the administrators fail to show a reasonable standard of care. Ignoring the complaint would constitute a clear failure to show reasonable care. Third, the breach must be directly related to the harm claimed. Recently, the term sick building syndrome has come to mean that a building is causing health problems, and the source is unknown.[2]

In the court case of Dean H.M. Chenensky, et al vs. Glenwood Management Corp., et al, there is a pending lawsuit involving $180 million regarding mold exposures. In another case, Robert E. Coiro, et al vs. Dormitory Authority of the State of

New York, the plaintiffs are seeking $65 million. Other suits involving various indoor air quality allegations are on-going.

Thus, the driving force in indoor air quality investigations is becoming litigation. The cost of a thorough indoor air quality investigation is small as compared to the cost of litigation.

DIFFERENCES IN HEALTH EFFECTS

The health effects of poor indoor air quality are dependent upon several factors. Relevant considerations when determining potential health effects on a population are the effect of each air contaminant, concentration, duration of exposure, and individual sensitivity.

The air contaminant may be an allergen, or it may be a carcinogenic chemical. The allergen will cause an immediate reaction with minimal long term effects. A carcinogenic chemical may not have any warning signs of exposure but may cause cancer years after exposure. It may be an irritant with passing health effects, or it may be a sensitizing chemical (e.g., isocyanates) whereby future exposures may result in an extreme immune response. Indoor air generally consists of a complex medley of substances that may have one or a combination of effects, and those substances that have the same health effect may not singularly cause health problems, whereas two different substances (e.g., irritants) may significantly impact human health when present at the same time.

Proper diagnosis is of course dependent upon proper identification of all contributing components. Then, once the contaminants have been identified, concentration should be ascertained.

Although there are known concentrations for many air contaminants at which well-defined health effects become evident, exposure levels defining the more subtle health effects are not as well researched. Furthermore, of the estimated 100,000 toxic substances to which building occupants are potentially exposed, fewer than 400 recommended exposure limits exist for industrial chemicals. OSHA regulates industry, and EPA regulates outdoor ambient air quality. Presently, no regulatory agencies control indoor air quality exposures limits.

Exposure duration is of particular concern in assessing indoor air quality exposures. In office buildings, exposures are generally 8 to 10 hours a day, 5 days a week. In residential structures, exposures may be up to 24 hours a day, 7 days a week. As some substances build up in the body over time, 24-hour exposures may result in an accumulation with the subsequent impact on health effects. Thus, the impact of a given concentration of air contaminant is less in office buildings than in residences. Other areas that should be considered potential long duration exposures include hospital patient rooms, hotels, mental wards, and prison cells.

Individual sensitivity contributes a huge variable to the combination of factors affecting the health of building occupants. Infants, elderly people, and sickly people are the most vulnerable to the health effects of air contaminants. Immune-suppressed individuals (e.g., AIDS patients and organ transplant recipients) and those with genetic diseases (e.g., *Lupus erythematosus*) are particularly sensitive to common molds. Individuals who drink alcohol in excess are more susceptible to air contaminants that may affect the liver. People with dry skin are more susceptible to further drying and skin penetration by chemicals. Those who smoke tobacco products have diminished body defense mechanisms. Certain medications enhance the effect of environmental exposures. Individuals with predisposed conditions (e.g., lungs damaged by fire) may have a heightened response to air contaminants.

A MISGUIDED PREMISE

As they compare the indoor to industrial environments, traditional industrial hygienists see the good, the bad, and the ugly. The indoor air quality environment is the good, and industrial environments are the bad and ugly. This is a misguided premise that requires comment.

Indoor air quality exposures involve multiple exposures to unknown substances in enclosed environments often with minimal fresh air, no exposure duration limits, and a wide range of individual susceptibility. Industrial exposures involve limited exposures to known chemicals in work environments with local exhaust ventilation, limited exposure duration, and healthy adults. The indoor environment does not begin to compare to the dirt-and-grime of industry. Yet, clearly there are differences.

A tight building has multiple factors that can result in sick building syndrome. An enclosed environment does not mean clean air.

REGULATORY LIMITS AND GUIDELINES

Presently, federal regulatory limits for indoor air quality are only implied. They are implied by U.S. government directives, the EPA Ambient Air Quality Standards, and OSHA standards. The regulated limits have been found inadequate in most cases involving indoor air quality.

In an effort to stem the tide of indoor air quality health complaints, various recognized public contributors have made an effort to recommend guidelines. Regulatory standards are mandated and guidelines are recommended. The recommended guidelines are more apt to appropriately address indoor air quality problems than the regulated standards.

Contributors to the guidelines consist of medical professionals, scientists, and engineers from many different disciplines. Major contributors are the American Conference of Governmental Industrial Hygienists (ACGIH) and the American Society of Heating, Refrigerating and Air-Conditioning Engineers (ASHRAE). ASHRAE has, in turn, credited all of the aforementioned contributors and some others including the World Health Organization (WHO). Although some of these standards are similar, most are not.

U.S. Government Directives

A limited number of federal agencies have been given directives to consider indoor air quality in their standards. In 1994, the Department of Energy was directed to consider the impact of energy-efficient options on habitability and persons and to achieve a balance between a healthy environment and energy conservation.[3] In 1997, the Department of Housing and Urban Development promulgated standards for the construction and safety of manufactured housing that includes features related to indoor air quality.[4]

EPA Ambient Air Quality Standards

The EPA air quality focus is to protect human health outdoors in the ambient air. The principal program that may be of some value to the reader is the National Ambient Air Quality Standard. The intent of this standard is to control emissions of six pollutants and their precursors when released in large quantities (e.g., vehicle exhausts and industrial emissions). This standard may be applied in indoor air quality investigations where the outside air may potentially contribute to exposure levels indoors such as in large non-attainment cities. Non-attainment means the city does not comply with one or a combination of the air quality standards as set forth in Table 1.1. Where exceeded outside, the National Ambient Air Quality Standards are likely to be exceeded indoors as well.

Table 1.1 National Ambient Air Quality Standards for Outdoor Air

	Long Term			Short Term		
	Concentration		Averaging	Concentration		Averaging
Contaminant	µg/m^3	ppm		µg/m^3	ppm	
Sulfur dioxide	80	0.03	1 year	365	0.14[A]	24 years
Total particulate	75[B]	–	1 year	260	–	24 years
Carbon monoxide				40,000[A]	35[A]	1 hour
				10,000[A]	9[A]	8 hours
Oxidants (ozone)				235	0.12[C]	1 hour
Nitrogen dioxide	100	0.055	1 year			
Lead	1.5	—	3 months[D]			

[A] Not to be exceeded more than once per year.

[B] Arithmetic mean.

[C] Standard attained when expected number of days per calendar year with maximum hourly average concentrations above 0.12 is equal to or less than 1 as determined by Appendix H to subchapter C, 40 CFR 50.

[D] Three-month period is a calendar quarter.

OSHA Workplace Standards

OSHA claims jurisdiction over all workplace environments. The workplace includes indoor air quality exposures in office buildings as well as in industry and construction. Yet, when it comes to indoor air quality, OSHA capabilities are limited in that the contaminants must be known and permissible exposure limits are based on outdated limits published by ACGIH in 1968.

Not only are most OSHA limits easily attained in indoor air quality investigations, but there are no provisions for low level irritants, molds, and allergens. Those investigators who do insist on applying the OSHA standards in office environments will generally find a dead end street. These same investigators often state that the OSHA standards have been met, so there must not be a problem. In a building where 80 percent of the occupants have health complaints, a statement that infers the only problem is mass hysteria will most assuredly find the investigator's credibility questioned.

The original OSHA exposure limits were derived from the 1968 ACGIH recommendations. Limits for only a handful of chemical contaminants (e.g., asbestos and benzene) have since been updated. For this reason, most industrial hygienists consider OSHA limits outdated and opt to use the ACGIH guidelines. Although backed up by the force of federal law, the OSHA limits are rarely exceeded in office environments where one or more of the contaminants have been properly identified. The complex nature of indoor air quality is not supported by OSHA limits.

ACGIH Workplace Guidelines

The ACGIH is a professional society of scientists who annually review and recommend guidelines to industrial hygienists for use in the assessment of occupational workplace exposures.

> *The ACGIH limits are intended for use in the practice of industrial hygiene as guidelines or recommendations in the control of potential workplace health hazards and for no other use . . . These limits are not fine lines between safe and dangerous concentrations nor are they a relative index of toxicity . . . a small percentage of workers may experience discomfort from some substances at concentrations at or below the threshold limit, and a smaller percentage may be affected more seriously by aggravation of a pre-existing condition or by development of an occupational illness.*

There are around 400 chemicals listed with recommended 15-minute and 8-hour exposure limits. These guidelines were created to address exposures in the workplace. Occupational exposures are generally limited to 8-hour exposure durations for healthy adults between the ages of 18 and 65. Thus, the ACGIH exposure guidelines do not apply to residential exposures where the exposure parameters differ. ASHRAE has addressed this consideration.

ASHRAE Criteria for Public Buildings[5]

In 1981, ASHRAE introduced a revised mechanical ventilation standard that is now referred to as the "Ventilation for Acceptable Indoor Air Quality Standard." ASHRAE developed and evolved consensus guidelines to address indoor air quality in public buildings.

Consensus is defined as "substantial agreement reached by concerned interests according to the judgment of a duly appointed authority, after a concerted attempt at resolving objections. Consensus implies much more than the concept of a simple majority but not necessarily unanimity." This definition is according to the American National Standards Institute (ANSI) of which ASHRAE is a member.[6]

The purpose of the standard is to "specify minimum ventilation rates and indoor air quality that will be acceptable to human occupants and are intended to avoid adverse health effects."[5] The health effects information and acceptable exposure limits rely on recognized authorities and their recommendations. Thus, the ASHRAE standard on "Ventilation for Acceptable Indoor Air Quality" has become the most commonly cited guideline for the investigating indoor air quality in commercial and institutional facilities.

The standard is intended to provide ventilation design and maintenance practices for air handling systems except where greater design specifications apply. It should be noted that ASHRAE has issued a disclaimer that "acceptable indoor air quality may not be achieved in all buildings meeting the requirements of this standard."[5]

Although not part of the actual standard itself, indoor air quality contaminant exposure limits are discussed extensively in Appendix C, ASHRAE 62-1999, "Guidance for the Establishment of Air Quality Criteria for the Indoor Environment." Each of the components is discussed herein.

ACGIH Guidelines Revisited

ASHRAE recommends the investigator start with an acceptable limit of one tenth of the ACGIH TLVs.

A concentration of 1/10 TLV would not produce complaints in non-industrial population(s) in residential, office, school, or other similar environments. The 1/10 TLV may not provide an environment satisfactory to individuals who are extremely sensitive to an irritant . . . Where standards or guidelines do not exist, expert help should be sought in evaluating what level of such a chemical or combination of chemicals would be acceptable. [7]

EPA and State Environmental Standards Revisited

The outdoor regulatory standards in Table C-1, "Standards Applicable in the United States for Common Indoor Air Pollutants," provide a list of acceptable concentrations in the United States for common outdoor air contaminants. The list includes federal and some state standards.

These standards may be applicable where the quality of the outdoor air is a potential contributor or suspect source in an indoor air quality investigation. Most would normally be found only in areas near industrial air emissions, but sometimes these limits are in residential environments.

Other Contributors

The indoor regulatory standards in Table C-1 are an abbreviated list of "Standards Applicable in the United States for Common Indoor Air Pollutants" that are regulated by the Consumer Product Safety Commission (e.g., asbestos and lead) and Housing and Urban Development (e.g., formaldehyde). These are applicable in all occupied spaces.

The indoor guidelines in Table C-2 are an abbreviated list of "Guidelines Applicable in the United States for Common Indoor Air Pollutants" that are recommended by the National Academy of Sciences (e.g., chlordane) and EPA (e.g., radon). These are recommended guidelines in residential environments.

The indoor guidelines in Table C-3 provide a "Summary of Canadian Exposure Guidelines for Residential Indoor Air Quality." Acceptable ranges are provided for eight air contaminants (e.g., 0.10 ppm exposure action level for formaldehyde) and for water vapor (e.g., 30-80% R.H. in the summer).

The indoor guidelines in Table C-4 provide the "WHO Working Group Consensus of Concern About Indoor Air Pollutants at 1984 Levels of Knowledge." This contains a listing of 22 commonly identified pollutants, concentrations reported, concentrations of limited or no concern, concentration of concern, and remarks. An example follows:

Pollutant	*Concentrations reported*	*Concentrations of limited or no concern*	*Concentration of concern*	*Remarks*
formaldehyde	*0.05-2 mg/m^3 (0.04-1.62 ppm)*	*<0.06 mg/m^3 (<0.05 ppm)*	*>0.12 mg/m^3 (>0.1 ppm)*	*long- and short-term*

ASHRAE Criteria for Residences

As of January 2001, ASHRAE has been in the process of developing criteria for residential environments. The standard is referred to as Ventilation for Acceptable Indoor Air Quality in Low-Rise Residential Buildings. The proposed standard is being developed to set guidelines to achieve acceptable indoor air quality for homes by ensuring minimum ventilation.

In addition to the whole-house mechanical ventilation, the proposed guidelines are to include:

- Source control of moisture and other specific improvements through the use of exhaust fans
- Local ventilation in wet rooms to remove odor and moisture
- Carbon monoxide detectors
- Criteria to minimize back-drafting and other combustion-related contaminants
- Provision to reduce contamination from attached garages
- Guidance on how to select, install, and operate systems

SUMMARY

Tight building syndrome and sick building syndrome have become household phrases. As indoor air quality complaints escalate, ignored health complaints in public buildings are becoming the rationale for lawsuits, and homeowners are living in fear. In an effort to stem the tide, environmental professionals are developing guidelines and recommendations that specifically address indoor air quality.

Indoor air quality investigations have yet to be standardized, regulated, or managed with consistency. Thus, those performing these investigations must develop a strong knowledge base and actively pursue each new case with the enquiring mind of a detective.

REFERENCES

1. *USA Today*. "Struggling to diagnose sick buildings." July 20, 2000. (usatoday.com/life/health/general/lhgen253.htm)
2. Hays, Larry. Lawsuits in the Air. American School & University. June 2000, 72(10):35.
3. Energy Conservation and Production Act, Pub. L. No. 94-385, 90 Stat. 1125 (1976); 12 U.S.C. section 1701z-8 (1994); 15 U.S.C. section 787 (1994); 42 U.S.C. sections 787-790 (1994); 42 U.S.C. sections 6801-6892 (1994).
4. 42 U.S.C. section 6851 (1997).

5. ASHRAE Standards Committee. Ventilation for Acceptable Indoor Air Quality. ASHRAE Publications, Atlanta. ASHRAE 62-1999.
6. ASHRAE Standards Committee. Ventilation for Acceptable Indoor Air Quality. ASHRAE Publications, Atlanta. ASHRAE 62-1989.
7. ASHRAE Standards Committee. Ventilation for Acceptable Indoor Air Quality: Guidance for the Establishment of Air Quality Criteria for the Indoor Environment. ASHRAE Publications, Atlanta. ASHRAE 62-1900, Appendix C. p. 17.

Chapter 2

PRELIMINARY INVESTIGATION

In accordance with a study conducted between 1971 and 1988, 34 percent of all sick building syndrome buildings assessed by NIOSH were found to be associated with indoor air contaminants, outdoor air contaminants, building materials, or microbes. The remaining 66 percent of the buildings had no sources identified. Either inadequate ventilation was implicated as the cause, or the source and cause were unknown.[1] These statistics are presented in Figure 2.1. Some environmental professionals feel that those buildings where there are no identified sources are the result of perceived poor indoor air quality and subsequent psychological effects on the health of the building occupants.[2] For this reason, it is important to properly investigate health complaints.

Although it has been the experience of the author that there is a typical complaint rate of 8 to 12 percent in office buildings, one publication claims a dissatisfication rate of 20 percent.[3] The most common complaints are that of temperature and humidity extremes, odors (e.g., cafeteria food and cigarette odors), noise (e.g., copy machine operation), and inadequate lighting. In some instances where these complaints are not addressed immediately, the occupants may develop stress-related problems (e.g., headaches).

In a survey funded by the Environmental Protection Agency, 20 non-problem buildings around the United States were surveyed in order to develop a baseline of complaints in structures not identified as sick buildings. Data regarding specific complaints arising from buildings with no indoor air quality problems was as follows:

- Headaches—19%
- Eye irritation—16%
- Fatigue—15%
- Sinus congestion—12%

As they may contribute to increased occupant health complaints, the investigator should be aware of these influences when assessing sick building syndrome.

Sick building syndrome occurs when health concerns increase significantly over that which is normal or when health complaints are more apparent and involve more than one person. Typically, most cases involve subtle symptoms that cannot be medically diagnosed.

Where the health symptoms are more obvious, there is generally a sudden release of a toxic substance into a limited area. The release, occasionally referred to as the smoking gun, may or may not involve an entire building. For example, a prankster released mace into the air in a grocery store. Eye irritation and breathing difficulties with a sense of suffocation resulted in an evacuation of the entire store. This latter scenario led to a series of events that culminated with carbon monoxide exposures from the operating fire truck at the front of the store where the patrons and store employees awaited further directions.

Keep a perspective of each new situation. Although there may be a subtle change in the number of health complaints within a building (e.g., 20 percent complaints), an isolated area within a building may experience as much as 100 percent complaints. For instance, in a building with 800 occupants, 40 complaints from one section of the building may be interpreted in two different ways. Eighty complaints in a building of 800 occupants is a 5 percent complaint rate, low by most people's standards. Yet, if 40 of the complaints all occur within the same area of the building with the number representing 100 percent of the occupants in that area, proper identification of the problem area will prevent an oversight that would otherwise occur if the entire building was assessed as a single area.

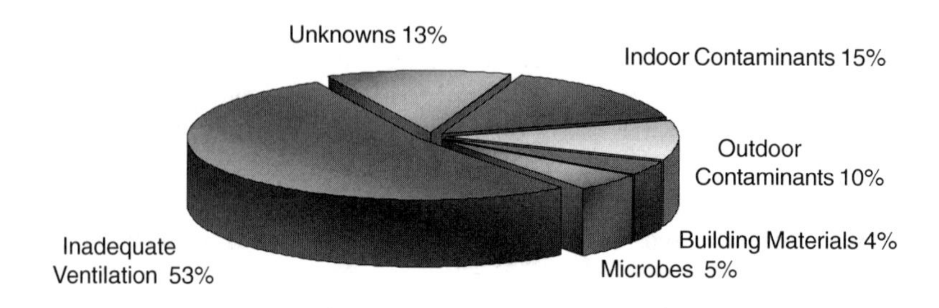

Figure 2.1 Source of Complaints in Buildings Assessed by NIOSH Study

As it is the objective of each investigator to attain good indoor air quality, the good as well as bad deserves mention. Good indoor air generally includes:

- Acceptable temperature and relative humidity
- Adequate fresh air and air movement
- Lack of offensive odors (e.g., heavy perfumes and mildew)
- Lack of significant levels of air contaminants

The result of good indoor air quality is improved occupant health, comfort, and productivity. This is certainly incentive for facilities managers to develop a proactive response program for occupant complaints and to seek the assistance of an investigator in response to identified sick building syndrome.

As part of a proactive response program, many of the steps provided herein can be performed by an indoor air quality team. All the steps should be considered tools only. The investigator may choose to overlook some steps in one situation and follow them in their entirety in another. In other instances, the investigator may not need to proceed through all of the steps. For example, an investigation initiated by complaints of headache and gasoline odors may require a quick investigation of the building to locate sources of fuel combustion by-products.

DOCUMENTS REVIEW

Obtain and review the building layout, mechanical blueprints (if available), an inventory of activities, an inventory of known chemicals, custodial activities, and pesticide treatment activities. Additionally, some investigators attempt to obtain full architectural plans, specifications, submittals, sheet metal drawings, commissioning reports, adjusting and balancing reports, inspection records, and operating manuals.

The building layout is a must have, particularly in public buildings. It may be in blueprint form, or the schematic may be a fire exit plan. The latter is more likely to be available and updated. As-built drawings are rarely kept up to date.

Mechanical blueprints for a building are rarely available, especially for older buildings. If they are present, however, they are often outdated. This document is one of the single most important documents the investigator will require. An alternative backup is to have someone knowledgeable in the mechanical operation of the building (e.g., building engineer or maintenance) sketch which air handler supplies what areas. This same person should also be asked to update altered as-built drawings to the best of their ability.

Identify each area by activity(ies). Get specific information. For instance, activities may include word processing and filing carbonless copy paper, legal casework, operating a blueprint copy machine, gluing/paste-up work, and word processing. The facilities personnel may or may not be able to provide this information. If not available, the activity information can be collected during the actual walk-through.

An inventory of chemicals should be collected wherein the potential for chemical exposures exists. In most public and institutional buildings, chemical information is sketchy at best. They may either generalize substances as custodial cleaning fluid or copy toner, or they may be able to provide material safety data sheets for all chemicals housed on the premises. The latter is the least likely to occur.

Relevant to the investigator during the walk-through and while developing an air sampling strategy, custodial activities and schedules should be obtained along with the type of supplies used. This may take some research on the behalf of the facilities manager or may require the investigator to schedule interviews with the custodial personnel.

Custodial activities are an often over looked contribution to indoor air quality, because they generally operate after hours. Yet, their activities have been found to impact the indoor air quality significantly. In one case, the custodial personnel used feather dusters in the office spaces and emptied their vacuum waste while in the office spaces (and wearing a paper respirator). In the morning, the office employees complained of visible dust in the light streaming through the windows.

Pesticide treatment activities are generally out of sight and out of mind. Although the scheduled peak treatment periods may coincide with complaints, the health effects of pesticides may be overlooked. In one case, spraying for a cockroach infestation resulted in airborne allergenic parts and pieces, a situation that could well have been avoided with roach motels.

BUILDING WALK-THROUGH

The intent of the walk-through is to acquire an overview of the building and occupant activities. A residential walk-through is considerably less complicated than one involving public buildings.

In public buildings, an initial walk-through should be planned and coordinated. Schedule the walk-through to include the occupancy periods and normal building operation. This accomplishes two things. It shows response to and concern for the occupant complaints, and the building can be assessed as it is when fully operational. The investigator is less likely to overlook relevant considerations during peak complaint periods.

In the interest of time, some investigators scale down the initial walk-through considerably on the first visit by performing a documents review and assessing questionnaires. With this information, the investigator can then develop an air sampling strategy and complete the walk-through at a later date while collecting air samples. Each investigator should be flexible enough to revise the approach, as scenarios and conditions differ from one investigative building to the next.

Occupied Areas

The investigator should have a set of floor plans (or a schematic) with all relevant information assessed during the documentation review. Several color markers may be helpful along with a pen, paper, clipboard (or binder), and flashlight. A building representative familiar with the air handling system should accompany the investigator with a set of keys and tools, and a ladder should be accessible as well.

The general condition of the occupied spaces in a building should be assessed. Some items to look for include, but are not limited to, the following:

- Odors
- Lint and/or soiling on carpets
- Dirt on sheet vinyl and floor tiles
- Dust on surfaces (e.g., desks and ledges)
- Water damage stains (e.g., ceiling tiles with tea-like stains)
- Suspended dust in air (e.g., observed in light)
- Moisture collecting on surfaces (e.g., condensation on windows)
- Dirt and debris around the air diffusers
- Cloth verses vinyl upholstery

- Presence of homeowner air purifiers
- Cleanliness of kitchen/food areas
- Rotting food in office spaces and trash receptacles
- Presence of plants
- Wall penetrations
- Peeling paint and vinyl wallpaper
- Storage of chemicals

Continue to add to the list as the situation dictates. Always keep in mind, "If it looks out of place, it probably is!" Ignoring this old adage could culminate in an oversight. Even if an unexplained sense of something not right should occur, attempt to identify the reason.

Air Handling System

An investigation of the air handling system generally requires some basic knowledge beyond the scope of this book. Such an assessment may require the assistance of a mechanical engineer or an industrial hygienist knowledgeable in HVAC systems.

Although the investigator may not be knowledgeable, air handling systems are one of the single most important contributors and/or solutions to indoor air quality problems. Some of the more basic items to look for in a centralized air handling unit include:

- Rust, damage, and water leaks around the exterior of the units
- Rust, dirt, and debris build-up inside the units
- Condition and amount of water in the drip pans
- Appearance of a slime or fuzzy growth
- Type, condition, and location of air filters (e.g., a very dirty, shredded, polyfiber filter prior to the cooling coils)
- Fit of the filters and visible space gaps between filter banks
- Measure and calculate the amount of fresh air intake in terms of cubic feet per minute per person (CFM)
- Dirt, damage, and moisture build-up in the duct liner immediately after the cooling coils
- Condition of the fan belt and motor
- Debris, dirt, and damage to the fan blades (e.g., a cardboard box blocking the opening)
- Biocide treatment used properly in drip pans
- Air deodorants in air supply systems
- Water drain not plugged
- Tools and equipment stored in an air handler
- Correct operation of computerized systems (e.g., computer records indicate the vents are opening when observation indicates otherwise)

Between the main air handling units and the occupied spaces, many systems will have additional conditioning/reconditioning units. As they are generally in the ceiling spaces, these units are a little more difficult to access and will require a ladder. All the above items should be included in an assessment of the support units.

In the occupied spaces, the air supply can be measured in terms of CFM delivery and the measurements compared with the specifications. In some instances, building occupants request their air supply be damped down. There is no air movement, and these same individuals are complaining. Sometimes occupants will get creative and jury rig a cardboard diverter so the air will not blow directly onto them. In the absence of air supply measuring equipment, the investigator should minimally make these observations.

In cases requiring negative or positive pressure in a building, pressure differential should be measured. Once again this may require some outside assistance. In hospitals, operating rooms require positive pressure and TB patient rooms require negative pressure. In buildings with considerable infusion of outside air through the wall spaces, investigators deeply entrenched in mold remediation are finding that delivery of positive pressure alleviates the mold problem considerably. Thus, positive/negative pressure measurements may be beneficial in some instances.

Bathroom Air Exhaust

Bathrooms have their own exhausts, and they are frequently poorly maintained. Sometimes the exhaust fan has ceased to function, the exhaust vent has been closed, or the vents are covered with debris. When the bathroom exhaust is associated with poor indoor air quality, building occupants might comment, "The air smells like a toilet." In some instances, the exhaust air from the bathrooms gets captured and entrained in the building air supply system, or the bathroom air is drawn out into the hallways and vicinity of the bathrooms where the bathrooms are not adequately exhausted.

A quick easy method to inspect the bathroom exhausts is to observe movement of dust on vents or to hold toilet tissue to the vent. Observe the direction of movement. If the tissue hangs down, it is a safe bet that the air is not being exhausted.

Occupant Activities

Observe special activities within a building—ordinary and out of the ordinary. Beyond the facilities personnel interviews, the investigator may observe activities that were not brought up in the discussion. This may involve renovation projects, or commercial and industrial activities. The investigator should become familiar with for all activities being conducted in the building, and then determine chemical usage, direction of air movement, and local exhaust locations.

INTERVIEWS WITH FACILITIES PERSONNEL

During the walk-through, the maintenance and custodial personnel should be interviewed. Keep records of all these conversations. Information that may not seem relevant at the time may become important later.

Maintenance Staff

The maintenance worker that is most familiar with the air handling system should be interviewed. In smaller facilities, this may be one individual who manages everything from nasty toilet odors and bats to complex electrical problems. In the larger facilities, someone may be dedicated to HVAC management. Whichever is the case, seek assistance from the person most knowledgeable in the system, and obtain the following information:

- Location of fresh air intake, local exhaust units, bathroom vents, and cooling tower(s) (if not indicated on the blueprints/floor plans)
- Placement of air filters (e.g., air handling unit or air return grid)
- Frequency of filter changes in the air handling system
- Special problems with the units
- Rationale and usage of air deodorants and biocides in the air handling system
- Typical occupant complaints and maintenance response
- Chemicals used frequently and associated health effects
- Location of stored chemicals
- Management of hazardous materials and/or waste
- Methods and associated complaints when accessing work above ceiling tiles
- Recent renovation and/or construction activities
- Methods and schedule for pesticide control

Although a wealthy resource of information, maintenance personnel have a tendency to be overlooked all too often. Give them the opportunity to voice their impressions and special observations that may be relevant to the investigation. This encourages cooperation and participation of behind the scene contributors, and there may be some enlightened information that would otherwise have been unacknowledged.

Custodial Staff

Another overlooked contributor to the indoor air quality is the custodial staff. Out of sight, out of mind. However, custodial activities can greatly impact the air quality of the building. Be forewarned that many are on contract and have a tendency to feel defensive when quarried as to their activities. You may find someone who speaks impeccable English lapse into, "I don't speak English." The custodial staff interview should include:

- Cleaning schedule for the various areas
- Type of vacuum cleaner
- Frequency and thoroughness of vacuuming
- Dusting procedures, frequency, and thoroughness
- Chemicals used frequently and associated health effects
- Location of stored chemical
- Waste management procedures and frequency

Encourage further input of information and avoid criticizing their procedures. The investigator's purpose is to collect information. Recommended procedural changes can come at a later date.

OBSERVATION OF SURROUNDING AREAS

Observe activities outside the building. Traffic movement and peak road usage periods may impact the indoor air quality. Identify areas where automobile, truck, and forklift exhausts may enter the building. The building air intake or wall penetrations might be so located that vehicular exhaust contributes to the indoor air quality.

Roof and road asphalting are generally associated with building complaints. Proximity to the fresh air intake should be considered, and information regarding the approximate time of day and duration of the pot operation should be noted.

Take note of industrial and commercial activities—location of exhaust stacks, apparent visible emissions, and type of industry or commercial activity. Suspect air contaminants can sometimes be associated with known predicable environmental contribution by industry type. For instance, a common contributor in commercially zoned areas may be naphtha from a dry cleaning operation.

Whenever outside air is suspect, note prevailing wind direction, the relative location of the building fresh air intake, suspect point sources, and exhaust stacks. Observe all possible mechanical and environmental conditions that may exasperate or contribute to the health complaints within the building.

ASSESSING OCCUPANT COMPLAINTS

The reason for an indoor air quality investigation is occupant complaints. Complaints are either ongoing or the focus of a baseline study. For this reason, an assessment of occupant complaints is the most important consideration when conducting a preliminary building investigation.

As individual health complaints can be biased and ambiguous, some investigators choose to perform interviews only. Yet, interviews can be unwieldy in office buildings with high occupancy. Thus, in high occupancy buildings, the investigator should both administer questionnaires and conduct limited interviews. Follow all questionnaires with random interviews or with interviews of select respondents.

Upon completion of the questionnaires and interviews, the purpose is to isolate, identify, and define complaint and non-complaint areas. Non-complaint areas are frequently overlooked by many investigators, but confirmation and clarification of these areas is important for comparison air sampling.

Questionnaires

Indoor air quality questionnaires should be designed to minimize bias, maximize the response rate, and provide information that is useful to the investigator. This is a tall order, not easy to fill.

Types of Questionnaires

As the questionnaire is to be filled out by a biased building occupant, bias is impossible to eliminate entirely. The lingering question remains, "How do I minimize bias?" There is no consensus in the response.

EPA and NIOSH published a one-page, open-ended questionnaire and an occupant interview form. Responses to such a questionnaire may be as brief as "I'm always sick" to "My doctor says I have sick building syndrome," or it may include extensive details. When crunching numbers and assessing open-ended questionnaires, the investigator will often encounter difficulties weighing the responses.

Some investigators have developed their own questionnaires with a listing of specific complaints. Others develop these questionnaires with a response scale (e.g., always, often, sometimes, or rarely). Opponents to this approach state that symptom labels can be interpreted differently. For instance, one person may interpret shortness of breath to mean slow, labored breathing while another may feel it means rapid, shallow breathing. One way around this may be to define or clarify some of the symptoms and leave space for comments.

Questionnaire Response Rate

A 100 percent response rate is a pipe dream. No matter how well designed a questionnaire, there will be some that just refuse to or are unavailable to complete a questionnaire. In marketing surveys, a 20 percent response rate is considered fairly good, but this is unacceptable in indoor air quality investigations. The author generally receives an 80 percent or better response rate with questionnaires based on the information contained herein.

To better design a questionnaire, the reasons for failure to respond should be taken into consideration. Some of the reasons for not responding include:

- Too time consuming
- Too difficult to read

- Never received it
- Not present for the duration of the survey (e.g., out of office)
- Lost it
- Didn't remember to turn it in
- No complaints
- Fear of management reprisals

Each one of the above issues can be overcome. Create a one-page, easy-to-read form with check boxes and comment space for each section either on the front page or on the back of the one-page questionnaire.

Color-code the questionnaires on the basis of air handling unit zones. This serves a dual purpose. First, colored paper is harder to lose on a desktop of all white paper. Second, organizing the questionnaires is easier when they have been color-coded.

Give the occupants enough time to fill out the questionnaires, but not so much time as to allow them to forget and/or lose it. In offices where occupants frequently work outside the office, a week is generally enough time. In offices where occupants are present the entire day, a couple of hours may be sufficient. The timing should be worked out with management.

Either on a cover sheet or in the heading of the questionnaire explain the rationale for the questionnaire and the importance of having non-complaint respondents along with the others. The information may also be made confidential between the consultant and the occupant.

Provide a time and location for the completed questionnaires. This should be attainable and in close proximity to where the occupant is located and/or departs the building.

Informational Data

A well-written questionnaire can provide important or potentially informative information to the investigator. The components may include, but not be limited to, the following:

- Physical location in the building
- Comfort level (e.g., perceived temperature and humidity)
- Odors
- Health concerns thought to be associated with the building
- Onset of symptoms (e.g., approximated date)
- Occurrence of symptoms (e.g., early morning on a Monday)
- Pre-existing conditions that might be more adversely impacted by exposures while in the building
- Occurrence of symptom relief (e.g., two hours after leaving work)
- Observed unusual/suspicious activities or events

As data is more easily assessed relative to complaint areas, physical location is a must. If reluctant to fill in this information, the respondent should be reminded of its importance and, at a minimum, provide the general area. Yet, area generalities are often inadequate and difficult to pin down. In the same light, people who tend to not stay put in an office environment are difficult to assess. Generalities make the final assessment quite challenging.

Comfort level allows the respondent to complain without associated temperature and humidity concerns with health complaints. They are not one and the same, but comfort level may enhance or reduce perceived health problems.

Odors are very subjective. Heavy perfume to one may be an overpowering stink to another. The smell of asphalt may be described by one person as sweet and by another as a chemical smell. The latter description is common for unfamiliar odors. Whereas most people describe mold by-products as mildew, others describe it as dirty feet or like the interior of a cave. While it is difficult to interpret odor descriptions, the information can assist the inspector to gleam direction as to the potential source of health complaints.

The actual health complaints should be associated with the building, not outside activities. If providing a checklist format, offer symptoms that are easy to interpret. For instance, the symptom of congestion may be replaced with stuffy nose. Irritated skin may be replaced with dry, flaky skin or itchy, red bumps. In a listing, try to keep the description of symptoms abbreviated and to the point.

The date of onset of symptoms is difficult if not impossible to tie down. Unless they experience health effects the first time they walked into the building, people do not notice they are having health problems until long after the initial exposures. Then, most of the occupants will live with it until they hear others expressing concerns. On the other hand, if an approximate time period can be associated with occupant activities, building renovation, or scheduled work activities (e.g., pesticide application), a narrowing of the gap on source identification may occur.

Pre-existing health conditions and/or medications that might be adversely impacted by exposures in the building may or may not be understood by the respondent, but they can occasionally provide some valuable insights. For instance, someone who has anemia would be more susceptible to the health effects of carbon monoxide that others. Asthmatics will experience worsened conditions when exposed to excessive airborne allergens (e.g., dust mites).

In most indoor air quality situations involving allergens in a building, the occupant will experience relief within 2 hours of departure. If the exposure is to carbon monoxide, relief may not be apparent for a couple days. This information can be very helpful to the investigator.

Permit the occupant some space to speculate and provide personal insight as to the source of the problem. Occasionally, occupant observations provide tremendous insight. It has been to the amazement of the author that people given a chance to contribute, unbeknownst to them, had the answer all along.

Interviews

When not used as the only means for gathering complaint information, interviews are where the investigator can clarify information and gain further details not obtained from the questionnaires. Unless you have a preconceived concern or special issue, allow the interviewees to tell you in their own words about their concerns. Concern is a less intimidating term than complaints. Listen to what they say without a preconceived notion, and follow up with questions to get a clear focus on that individual's concerns. If not already covered, all the items discussed above should be discussed as well. Yet, a touch of reality, some interviewees will gladly talk all day. Set a time limit, and at some point, politely move on.

SUMMARY

Steps include a documents review, building walk-through, interview of facilities personnel, observation of the surrounding areas, and an assessment of occupant complaints. All steps may or may not be required in each different situation, and the components are subject to change likewise.

A sample of NIOSH/EPA checklists and forms is available for review in Table 2.1. This is merely one approach, not a must do. It is an approach the investigator can gleam ideas from and adapt.

Keep in mind that the procedures presented herein can and should be expanded or utilized in part. When performing a preliminary investigation, attempt to avoid preconceived notions and biases. Be prepared to seek that which is intuited. An open mind is the investigator's greatest tool.

REFERENCES

1. Seitz, T.A. Proceedings Indoor Air Quality International Symposium: The Practitioner's Approach to Indoor Air Quality Investigations. American Industrial Hygiene Association, Akron, OH (1989).
2. Hedge, Alan, Ph.D. Addressing the Psychological Aspects of Indoor Air Quality. (Paper) Cornell University, Ithaca, NY (1996).
3. U.S. Environmental Protection Agency and National Institute for Occupational Safety and Health. Building Air Quality: A Guide for Building Owners and Facility Managers. U.S. Government Printing Office, Washington, D.C. (Dec. 1991).

Table 2.1 Sample NIOSH/EPA Checklists and Forms for Preliminary Investigation

HVAC CHECKLIST

Building: ______________________ File Number: ______________________

Completed by: ______________ Title: __________ Data Checklist: __________

Component	NA	OK	Needs Attention	Comments
Outside Air Intake				
Bird Screen				
Outside Air Dampers				
Outdoor Air Quality				
Mixing Plenum				
Filters				
Spray Humidifiers or Air Washers				
Face and Bypass Dampers				
Cooling Coil				
Condensate Drip Pans				
Mist Eliminators				
Supply Fan Chambers				
Supply Fans				
Heating Coil				
Reheat Coils				
Steam Humidifier				
Supply Ductwork				
Pressurized Ceiling Supply Plenum				
Thermal Equipment (supply)				
VAV Box				
Thermostats				
Humidity Sensor				
Room Partitions				
Stairwells				
Return Air Plenum				
Duct Returns				
Return Fan Chambers				
Return Fans				
Exhaust Fans				
Toilet Exhausts				

Smoking Lounge Exhaust

Print Room Exhaust

Garage Ventilation

Mechanical Rooms

Preventative Maintenance

Boilers

Cooling Tower

Chillers

Chemical Inventory

The inventory should include chemicals stored or used in the building for cleaning, maintenance, operations, and pest control. If you have an MSDS (Material Safety Data Sheet) for the chemical, put a check mark in the right-hand column. If not, ask the chemical supplier to provide the MSDS, if one is available.

Date	Chemical/Brand Name	Use	Storage Location	MSDS on File

Occupant Diary

On the form below, please record each occasion when you experience a symptom of ill-health or discomfort that you think may be linked to an environmental condition in this building.

It is important that you record the time and date and your location within the building as accurately as possible, because that will help to identify conditions (e.g., equipment operation) that may be associated with your problem. Also, please try to describe the severity of your symptoms (e.g., mild, severe) and their duration (the length of time that may persist). Any other observations that you think may help in identifying the cause of the problem should be noted in the Comments column. Feel free to attach additional pages or use more than one line for each event if you need more room to record your observations.

Time/Date	Location	Symptom	Severity/ Duration	Comments

Excerpted from NIOSH/EPA: Building Air Quality.[3]

Chapter 3

THE HYPOTHESIS

The objective of an investigator is to identify and solve indoor air quality complaints in a way that prevents a recurrence, not create other problems. Although this may seem an overstatement of the obvious, poorly researched and executed investigations are consistently recurring.

In one situation, an investigator recommended increased fresh air intake. Health complaints from the occupants were reduced from 90 percent to 20 percent. Another investigator convinced the building management that the 20 percent could be improved with a building "burn out." The source of the complaints had never been identified, but the second investigator, without further investigation, recommended a procedure that worked in another building. Sounds good! It should work again. The burn out involved elevating temperatures over a long weekend and flushing the building with 100 percent makeup air. The result was 75 percent health complaints, a recurrence of problems.

Another situation involved elevated mold spores and no diagnostic evaluation to determine the source of amplification. The investigator speculated that the spores were growing in the perpetually damp, building crawl space. The recommendation was to use powerful exhaust fans to move air through the crawl space and dry out the soil. The result was air movement over the soil and being exhausted outside where it was picked up by the air intake and distributed throughout the building. Health complaints worsened. The reason—an untested hypothesis, insufficient diagnostic sampling, and recommendations based on unsubtantiated speculation.

A well-researched preliminary investigation is the best avenue to avoid some of these pitfalls. A strong foundation of information should result in a hypothesis. Avoid assumptions! Gather information, formulate a hypothesis, test the hypothesis, and make recommendations. The preliminary investigative process was discussed in the preceding chapter. Herein, we discuss putting the information in an easy-to-assess format and formulating a hypothesis.

INFORMATION REVIEW

An easy-to-access format of the voluminous amount of data possible in performing an indoor air quality investigation can be daunting at best. The author has reviewed many different formats and found an approach that works best in most instances. The methods provided are simplified and should be expanded to suit individual needs. These assessment methods are a working tool to be expanded and improved upon.

Building Assessment

The building assessment includes information gathered from the building walk-through, interviews of facilities personnel, and observations of surrounding areas. If not simplified, the information may become a paper morass, overwhelming and confusing.

One suggested approach wherever possible is to summarize information on the building blueprint or schematic. This may even be the ever-present fire evacuation plan. Either take notes on the drawing during the walk-through or summarize important information on the drawing after compiling all the data. In a publication regarding building assessment methods, "Building Air Quality," NIOSH/EPA suggest this technique for assessing pollutant pathways.[1]

Color markers, colored pens, overlays with indelible markers, and/or a coding system are recommended tools. Several sheets of the basic floor plans are also helpful. Summary comments may include, but not be limited to, the following:

- Air delivery zones, based on air handling units supplying air to that zone.
- Plumbing discrepancies (e.g., leaking pipes)
- Ventilation discrepancies (e.g., air supply vent closed) or observations (e.g., redirected air flow)
- Unspoken comments or fixtures regarding perceived air quality (e.g., air purifiers and pedestal fans)
- Chemical storage areas
- Location and type of equipment that may evolve air contaminants (e.g., copy machine)
- Water damaged structural components (e.g., water damaged ceiling tiles)
- Lifting/poorly adhered floor tiles and sheet vinyl
- Peeling paint and vinyl wallpaper
- Rusted metal structural components (e.g., air vents and window frames)
- Room(s) with carpeting
- Stains of flooring (e.g., burned floor tiles or stained carpeting)
- Bathroom exhaust discrepancies
- Location of elevators and other structures that may impact air movement (e.g., doors and stairwells)
- Plants
- Wall penetrations
- Areas where investigator, not occupants, noticed odors
- Location of bathrooms and custodial closets
- Commercial/industrial activities in the building
- Approximate direction of commercial/industrial activities outside the building
- Direction of prevailing winds
- Location of break room/kitchen
- Signs of infestations (e.g., cockroaches, rodents, bats, pigeons)
- Local exhaust ventilation

Figure 3.1 Sample Sick Building Field Notes Taken on Building Schematic

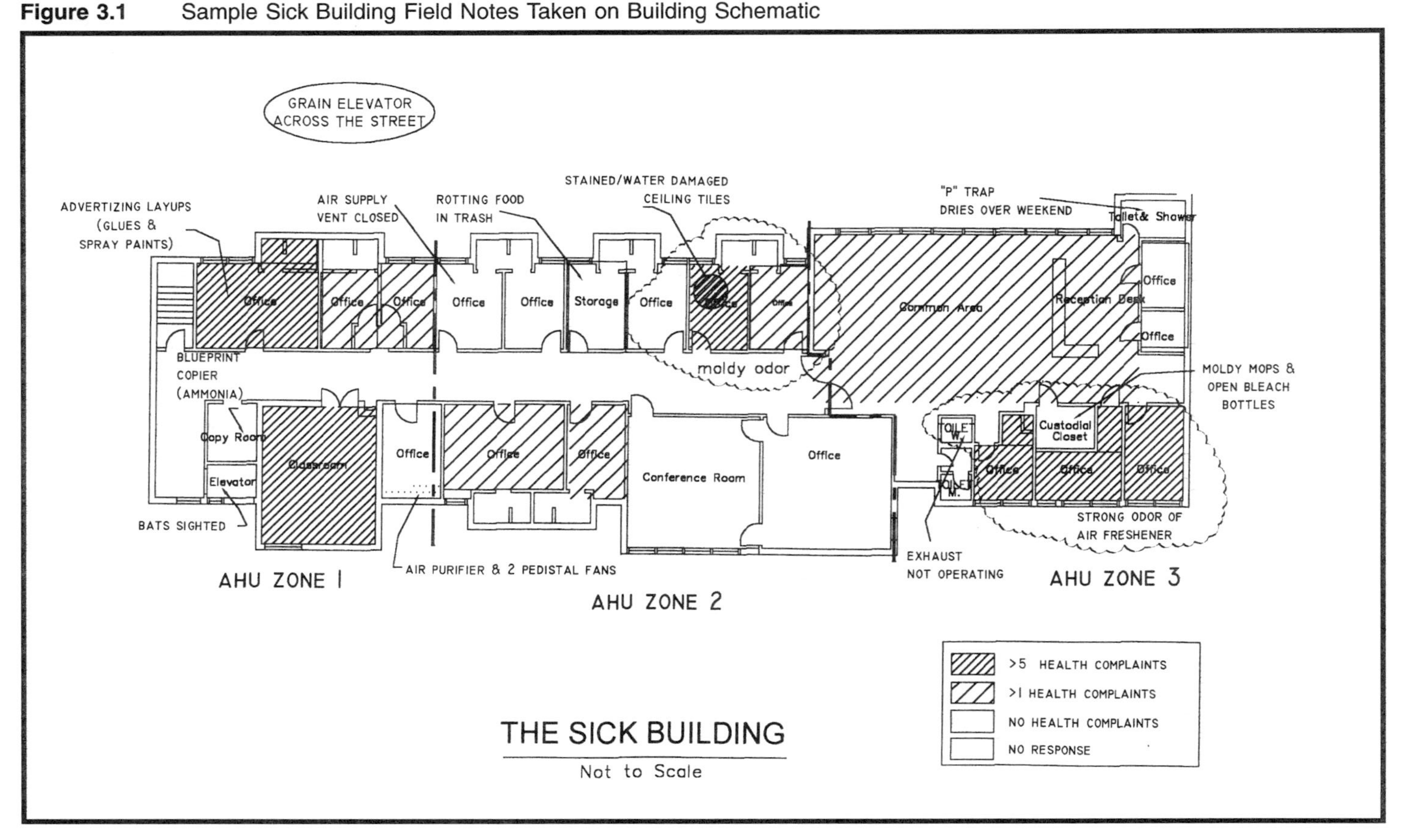

Comment about and enter information that does not seem relevant at the time. When assessing the relative locations of potential sources, discrepancies, and other observations as one, the investigator may be able to relate one previously disassociated item to another. See Figure 3.1.

Complaint Occupant

The single most important source of information is the occupant. With building complaint input, the investigator can define the complaint area(s) and determine the severity of the complaints as well as look for symptoms, occurrence of symptoms, and perceived associations. Problem and non-problem areas can be identified. Problem areas may be compared to the building assessment findings, and a potential causative agent can be projected. There is just one factor that complicates the process.

In a large occupancy building, the investigator must collect a large volume of information and reduce it to a manageable format. Color-coding questionnaires (e.g., coded air handling zones) is a good start. For confidentiality, each set should also be numbered, summarized, and filed by the investigator. The numbered and summarized information may then be installed into a spreadsheet and/or color-coded on the floor plans. It is best to make entries in both areas, and clarity is more apt to be gleamed from specific sites occupied by each of the respondents. Where there are no cubicle or office numbers, a little more creative coordination is required. For an example of the summary format, see Figure 3.2.

Upon completion of the complaint summary or interviews, the investigator should attempt to relate the symptoms to potential sources. This may involve a direct correlation, or the possibilities may be multiple.

In most sick building syndrome situations, the symptoms are allergenic in nature (Appendix A) and occur only when the occupant is in the building, and subside within a couple hours of departure from the building. As the recent trend has been to point an accusatory finger at molds, the full range of possibilities is frequently overlooked.

To a lesser extent, there are those sick building syndrome cases beyond the norm that involve a wider range of health effects (e.g., disease, febrile, flu-like symptoms, dermatitis, irritation, and systemic toxicity). These non-allergy health effects may be associated with a single substance, or they may be complicated by the effect of several substances. Several different agents may have the same health effect and result in an accumulation of different low-level exposures, or they may have different health effects contributing to the overall impact. For reference information concerning health effects, causative agents, and their occurrence, see Table 3.1.

Figure 3.2 Sample Sick Building Questionnaire Summary

Location	Headache	Fatigue	Eye Irritation	Itchy Throat	Sneezing	Coughing	Dry, Flaky Skin	Chest Tightness	Dizziness	Other	Comments	Occurrence
Zone 1												
1	X	-	X	-	-	-	-	-	X	-		no pattern; Dec '96
2	X	-	X	-	-	-	-	-	X	-	both; Mondays	Jun '97; water leaks
3	-	-	-	-	-	-	-	-	-	-	-	temp. extremes
4	X	X	-	-	-	-	X	-	X	ears ring	mostly at work	stuffy air; no movement
5	X	-	X	X	X	X	-	-	X	-	work; Mondays	May '99
6	X	X	X	X	X	X	X	X	X	nausea	work	Aug. '99; black in vents
7	X	X	-	-	-	X	X	-	X	-	work	something in vents
8	X	-	X	-	X	-	-	-	-	-	worse at work	-
9	(X)	-	-	-	-	-	-	-	-	-	work	May '98
10	(X)	-	-	-	-	-	-	-	-	-	both	-
	90%	30%	50%	20%	30%	30%	30%	10%	60%			
Zone 2												
1	-	-	-	-	-	-	X	-	-	-	work; morning	-
2	X	X	-	-	-	-	X	-	-	-	work; afternoon	late May
3	-	-	-	-	-	-	-	-	-	-	both	-
4	X	-	-	-	-	-	-	-	-	-	both	mildew in elevators
5	X	-	-	-	-	-	-	-	-	-	work	temp. extremes
6	-	-	-	-	-	-	-	-	-	-	-	-
7	-	-	-	-	(X)	-	-	-	-	-	work	-
8	-	-	-	-	(X)	-	-	-	-	-	work; morning	few months ago
9	-	-	-	-	-	-	-	-	-	-	-	-
10	-	-	-	-	(X)	-	-	-	-	-	work	black mold around vent
	30%	10%	0%	0%	30%	0%	20%	0%	0%			

Contribution by Omega Southwest Consulting, Inc. in Georgetown, TX.

Table 3.1 Reference Chapters for Relating Symptoms and Source Occurrence

Allergy Symptoms: associated with the building and symptoms subsides within a couple hours of departure

Chapter 4	Pollen and Spore Allergens
Chapter 5	Viable Microbial Allergens
Chapter 14	Forensics of Dust
Chapter 15	Animal Allergenic Dust

Disease: associated with a building and symptoms continue until the illness has run its course

Chapter 6	Pathogenic Microbes

Dermatitis: associated with the building or a work product and symptoms may improve upon removal or departure from the area but recovery may take a few days

Chapter 4	Pollen and Spore Allergens
Chapter 5	Viable Microbial Allergens
Chapter 6	Pathogenic Microbes
Chapter 7	Toxigenic Microbes
Chapter 8	Volatile Organic Compounds
Chapter 12	Formaldehyde
Chapter 14	Forensics of Dust
Chapter 15	Animal Allergenic Dust

Eye Irritation: associates with the building and subsides upon departure

Chapter 4	Pollen and Spore Allergens
Chapter 5	Viable Microbial Allergens
Chapter 7	Toxigenic Microbes
Chapter 8	Volatile Organic Compounds
Chapter 9	Mold Volatile Organic Compounds
Chapter 12	Formaldehyde
Chapter 14	Forensics of Dust
Chapter 15	Animal Allergenic Dust

Systemic Illness: associated with the building and may subside over time after departure

Chapter 5	Viable Microbial Allergens
Chapter 7	Toxigenic Microbes
Chapter 8	Volatile Organic Compounds
Chapter 9	Mold Volatile Organic Compounds
Chapter 12	Formaldehyde
Chapter 14	Forensics of Dust

HYPOTHESIS DEVELOPMENT

With the completed building and complaint assessments, the investigator can project possible scenarios of cause and effect. Write down all probable explanations for the complaints. Look at the non-complaint areas as well as the complaint areas. Look for links between the building components and complaints. Project sources of exposure and pathways.

One area may have several sources that may contribute to the overall health complaints. For instance, a carpet infested with dust mites and air supply vents disbursing mold spores may result in allergy symptoms. If the investigator theorizes only the mold spores, encounters moderate airborne exposures, and remediates the air supply vents, the contributing dust mites will be missed and remain uncorrected with a level of continued allergy symptoms.

On the other hand, different areas of one building may have different problems and each of these as complex as that discussed previously. Area 1 may have rodent allergens and airborne fiberglass from the air duct lining, and Area 2 may have leaking Freon and excessive levels of carbon monoxide. Symptoms will be different in these areas. If considered part of a composite, the two areas will be overlooked as one. The cause and effect may be overlooked entirely.

These scenarios are not unusual. The investigator should keep an open mind at all times. Develop a set of hypotheses. Consider the pathways, and test the hypotheses.

There are two ways to test a hypothesis. The easiest way is to project the source and associated solution. Where a hypothesis projects an inexpensive solution and sampling is expensive, the hypothesis may be tested by manipulating building conditions and/or the ventilation system. The source of complaints as indicated in Figure 2.1 in many indoor air quality investigations had been identified in 56 percent of the study cases as inadequate ventilation. It would be more appropriate to state that ventilation rectified the problem, not that it was the problem.

Sampling is a more direct approach to proving a hypothesis, the only method whereby the investigator can definitively clarify the source or sources of occupant complaints. As with a medical doctor who develops a hypothesis as to an illness, the only way he can definitively prove the cause of the illness is to perform a series of tests. To properly treat an illness, the doctor must identify the appropriate tests that will provide the much needed information. If the diagnosis is incorrect, treatment of symptoms may or may not work. It is, likewise, preferable in indoor air quality that the hypothesis be tested, proven, and acted upon appropriately.

THE PROACTIVE APPROACH

In order to minimize the sampling and effectively target potential sources of indoor air quality complaints, the investigator would be best served by having performed a baseline study of a building when there are no complaints. Many facilities managers and institutions are proceeding this way.

A proactive survey involves a scaled down variation of the more in-depth investigation that results from health complaints. The up front costs are greater, but the end result pays off in dividends.

Develop a plan for preventing poor air quality. This plan should include, but not be limited to, the following:

- A means whereby complaints are addressed in a timely fashion
- A guide for determination of an excessive number of complaints in a building and/or area of a building
- A means for maintaining records of building complaints, activities, and renovations
- A response action

Establish a baseline of occupant concerns and a building profile. A baseline taken in a healthy building will provide valuable information should the facilities manager or investigator determine that the building has an inordinately high number of health complaints.

Perform a limited amount of air sampling. This should be kept at a minimum while providing data regarding contaminants that may be reasonably implicated in future concerns.

Steps include a walk-through of a facility, an assessment of occupant complaints, identifying problem niches, assessing the building, assessing activities associated with the building, and compilation of all the data. Making sense of the mass of information collected is the investigator's greatest challenge.

BEYOND THE SCOPE

There are literally hundreds and thousands of substances that go beyond the scope of a single textbook. Other areas of expertise that may be required include medical physicians, industrial hygienists, toxicologists, and psychiatrists.

Medical Physicians

Medical physicians can address special situations involving individuals that are particularly susceptible to various environmental agents, and they can perform some speculative testing of occupants to determine susceptibility. An example of susceptible individuals is immune- suppressed patients, individuals that are anemic, and asthmatics. The boy that lived in a bubble had no immune system to speak of. Were he to be exposed to the same environment as others, he would surely die.

Prescription and over-the-counter drugs can also affect one's reaction to environmental influences. If while taking an allergy medication that causes one to become drowsy, an individual is exposed to low levels of a chemical narcotic, the effect can be enhanced.

Speculative testing is not the ideal approach to indoor air quality situations, but it is applicable in some situations, particularly where there are no known environmental sampling methods readily available for testing a hypothesis. In a hospital, a group of people experienced allergy symptoms when working in an area where latex gloves were frequently changed. Upon performing latex allergy testing on some of the people, the physician was able to confirm latex sensitivities in those with the greatest number of complaints.

Occasionally, a person claiming to have multiple chemical sensitivity may be one of the building occupants. The mere uncapping of a magic marker will illicit a reaction. Perfumes and deodorants are intolerable. They cannot go to a gasoline station for fuel without special respirators. If one of these individuals is a respondent to an indoor air quality questionnaire, the response will likely be dissimilar to the others. Generally, these people will have sought diagnosis by a physician.

Industrial Hygienists and Toxicologists

Industrial hygienists are trained and experienced in the anticipation, recognition, evaluation, and control of health hazards. Some of the less frequently encountered indoor air quality issues involve substances that are beyond the abilities of an untrained individual to assess. These substances include metals and pesticides/insecticides.

Some of the metal exposures that may be considered in indoor air quality are airborne lead, mercury, and arsenic. This may appear simple on direct analysis, but an assessment must be made on the basis of the form a metal is in and the environment. For example, *Scopulariopisis bevicaulis* growing on wet copper arsenite pigment wallpaper will produce trimethoxy arsenic vapor. Arsine gas is evolved when arsenic compounds react with acids.[2] These different forms of arsenic would require different sampling methodologies, and this applies likewise to the other metals.

Misapplied pesticides can have adverse, sometimes life threatening effects on individuals. One of the most commonly sampled pesticides indoors is the banned termiticide chlordane. Although banned from use in the United States, DDT is still encountered in the environment and continues to be used in third world countries.

Industrial hygienists and toxicologists stay current with publications on recent findings in research that is not covered by the media. Of import to indoor air quality is a recent hazard review on carbonless copy paper. It was found that some types of carbonless copy paper result in symptoms of skin, eyes, and respiratory tract irritation, allergic contact dermatitis, and some unclear systemic reactions. Although the chemical agent has yet to be identified, the relationship was indicated through epidemiological studies.

In another case, occupants complaining of dermatitis on the back of their thighs all sit on fabric upholstered chairs. The chairs are found to be allergen-free. Yet, it is noted that the custodial staff use a strong disinfectant for cleaning the chairs. This is a case of guilt by association.

Psychiatrists

This is the skeptic's corner, the last recourse for finding the cause of sick building syndrome. As cases go unsolved, investigators turn more and more to perceived indoor air quality on the basis of physical, psychological discomfort and odors. Then, too, there is mass psychogenic illness (MPI), or mass hysteria.

> *Mass psychogenic illness refers to "the collective occurrence of a set of physical symptoms and related beliefs among two or more individuals in the absence of any identifiable [cause]." Social psychological processes of contagion, where complaints and symptoms spread from person to person, and convergence, where groups of people develop similar symptoms at about the same time . . . Environmental events, like an unpleasant odor, can trigger contagion and convergence processes, and occupants who cannot readily identify what has triggered their symptoms often attribute these to any visible environmental changes, such as installation of a new carpet, or invisible agents, such as "mystery bugs." MPI symptoms include headache, nausea, weakness, dizziness, sleepiness, hyperventilation, fainting, and vomiting, and occasionally skin disorders and burning sensations in the throat and eyes.*[3]

SUMMARY

Upon completion of a preliminary investigation, the information should be compiled and simplified. Otherwise, the investigator may become overwhelmed and may not be able to see the forest through the trees.

The building and occupant assessments should be critically reviewed and a hypothesis developed. The hypothesis may involve one or several potential sources, and it may be complicated by the existence of one or several disassociated areas. Too often this point is overlooked. Recommendations to remediate one of several problems may result in unsatisfactory results. The objective is to open the door to all possibilities, to limit the amount of testing required to prove a hypothesis, and to appropriately target identifiable, quantifiable sources.

The most direct approach to proving a hypothesis is sampling. Sampling may involve one, or a combination of, screening, air sampling, and diagnostic tests. Yet, even the testing may result in incomplete, inconclusive information—if not performed properly. There have been instances whereby an untrained, uninformed investigator has taken expensive samples where strategy was not considered, sampling methodologies were inappropriate, and interpretation was limited. It is the purpose of this book to provide direction for sampling some of the more commonly encountered substances in sick building syndrome. See Table 3.1 for suggested links in getting direction for making sampling decisions.

REFERENCES

1. U.S. Environmental Protection Agency and National Institute for Occupational Safety and Health. Building Air Quality: A Guide for Building Owners and Facility Managers. U.S. Government Printing Office, Washington, D.C. (Dec. 1991). p. 70.
2. Kaye, Brian H. *Science and the Detective.* VCH, New York, NY (1995) p. 9.
3. Hedge, Alan, Ph.D. Addressing the Psychological Aspects of Indoor Air Quality. (Paper) Cornell University, Ithaca, NY (1996). p. 3.

Section II

BIOAEROSOLS

Chapter 4

POLLEN AND SPORE ALLERGENS

Except in the most restrictive of environments (e.g., an environmentally controlled, filtered bubble enclosure), allergens are everywhere, and the most commonly recognized allergens are pollen grains and fungal spores.

Pollen grains are the male reproductive cells that are dispersed by plants and carried by insects, animals, and wind to fertilize the female flower of like species. They are typically outdoor allergens but have on occasion been found to be problematic due to the capture and retention of the pollen within an air system.

Spores, as presented herein, are to include all forms of fungal spores (e.g., mold spores and mushroom basidiospores). Fungal spores are reported to affect greater than 20 percent of the adult population. It should be noted, however, that some bacteria may also produce spores, and these are discussed more fully in Chapter 5.

Pollen grains and spores must be airborne in order to cause respiratory allergy symptoms, and the total exposure to all of these will have a varying affect on those exposed. The higher the exposures, the greater the number of people affected. Their impact is irrespective of viability, or their ability to grow. Dead molds do not go away. They merely stop reproducing and growing. Mold spores persist—dead or alive!

OCCURRENCE OF POLLEN AND SPORE ALLERGENS

Pollen grains are the male reproductive cells that are dispersed by plants and carried by insects, animals, and wind to fertilize the female flower of like species. Those that are carried by insects and animals tend to be sticky, posses an elaborate exterior surface (e.g., spines and heavy ridges), and are large by comparison to the other pollen grains (e.g., up to 250 microns in size). On the other hand, those that are dispersed by wind tend to be non-sticky, smooth, light in density, and small in size (e.g., generally less than 50 microns). These reproductive cells are produced by weeds, grasses, and trees. There are over 350,000 species of plants. Plants are geographic, and pollen production (i.e., pollination) is seasonal.

Fungi include single celled yeast, filamentous molds, and multi-cellular mushrooms. Possessing a hard chitin or polysaccharide exterior covering, fungal spores are typically resistant to drying, heat, freezing, and some chemical agents.

General Information

Allergenic spores and pollen may be transported by high winds as far as 1,500 miles, and it is possible to find them 100 miles from their point of origin.[1] If one were to draw a contour map showing levels at various points from a source, it would be evident the highest concentrations are close to the source, diminishing with distance and impacted by wind direction, velocity, and volume of pollen produced at the source.

Small fungal colonies may discharge as many as thirty billion spores per day. Pollen grain discharges may be likewise remarkable with numbers reported as high as seven trillion pollen grains per tree on a season. See Table 4.1.

Attempts have been made to identify allergenic pollen types and the times of the year when their local presence is increased. Some highly allergenic individuals make decisions for relocation based on the prevalence of given allergens. Although Table 4.2 demonstrates an effort to categorize by state, the determinations are generalized and may not be representative of local areas within the regions mentioned.

The size, shape, and density of the airborne allergens affect their aerodynamic

Table 4.1 Pollen and Spore Single Source Discharge Rates

FUNGAL SPORES—one colony or growth unit	
Ganoderma applanatum	30 billion per day
Daldinia concentrica	100 million per day
Penicillium spp.	400 million per day
POLLEN GRAINS—one tree	
European Beech (*Fagus sylvaticus*)	409,000,000 per year
Sessile Oak (*Quercus petraea*)	654,400,000 per year
Spruce (*Picea abies*)	5,480,600,000 per year
Scotch Pine (*Pinus sylvestris*)	6,442,200,000 per year
Alder (*Alnus spp.*)	7,239,300,000 per year

Excerpted from *Sampling and Identifying Allergenic Pollens and Molds.* [2]

characteristics while the air humidity, wind direction, wind velocity, and obstructions affect their travel path as well as their distance. Temperature, soil types, and altitude may also impact the quantity of airborne allergens.

The size of fungal spores range from 1 to over 500 microns in diameter/length, but those that are typically airborne range in size from 1 to 60 microns. The *Cladosporium* mold spores typically range between 4 and 20 microns in length. *Alternaria* spores are around 30 microns in length (ranging from 8 to 500 microns), and *Aspergillus*/*Penicillium* spores are around 1 micron in diameter. It should be noted that some spore-producing bacteria are also on the order of 1 micron in size

Table 4.2 Plant Allergens by Region

NORTHERN WOODLAND
Trees (April – June) — birch
Fungi (June – October) — mushrooms and puffballs; watertight cabins and cottages tend to be moldy

EASTERN AGRICULTURAL
Trees (March – May) — ash, birch, box elder, elm, mulberry, oak, sycamore, and walnut
Grass (May – July)
Weeds (July – September) — hemp, goosefoot, and ragweed
Fungi (May – November)
Other — castor beans, cottonseed, and soybeans

SOUTHEASTERN COASTAL
Trees (February – April) — ash, elm, oak, pecan, and sycamore
Grass (February – October)
Weeds (July – October) — ragweed
Fungi (all year)

SOUTHERN FLORIDA
Trees (January – April) — oak
Grass (January – October)
Weeds (June – October) — ragweed
Fungi Indoors (all year)

GREAT PLAINS
Trees (February – April) — oak
Grass (April – September)
Weeds (July – October) — goosefoot, ragweed, and sage
Fungi (May – November)
Other — livestock dander, fertilizer dust, animal feed dust, grain, and storage dust

WESTERN MOUNTAIN
Trees (January – March) — mountain cedar
Grass (May – August)
Weeds (July – October) — goosefoot, ragweed, and sage

GREAT BASIN
Weeds (July – September) — goosefoot and sage

SOUTHWESTERN DESERT
Trees (January – April) — Arizona cypress, mountain cedar, and mulberry
Grass (March – October)
Weeds (April – September) — goosefoot and ragweed
Fungi (increased by use of evaporative cooling units in buildings)

CALIFORNIA LOWLAND
Grass (March – October)

Table 4.2 Plant Allergens by Region (continued)

NORTHWEST COASTAL
Grass (May – September)
Weeds (May – August) — goosefoot

ALASKA
Other — dog dander

HAWAII
Grass (all year)
Fungi Indoors and Outdoors (all year)

PUERTO RICO
Grass (all year)
Other — insect parts, bat droppings, and smoke of burning sugar cane (irritant or allergen unclear)

Excerpted from the U.S. Pollen Calendar.[1]

and may appear microscopically to be mold spores and cannot be differentiated without growing the spores in nutrient agar. See Figure 4.1 for differentiation between two molds of similar spore production.

Pollen grains are typically denser and, on the average, larger in size than the fungal spores. They range from 14 microns (for stinging nettle) to 250 microns in

Figure 4.1 Differentiation between molds starts with the microscopic appearance of colonies and spores. This example demonstrates the colony appearance of two different genera of fungi that have similar spores, both around 1 to 2 microns in diameter, which is also the same as that of the larger spore-forming bacteria. The above drawings are: (left) *Aspergillus spp.* and (right) *Penicillium spp.*

microns. Tree and weed pollen are the more variable. Most, however, fall between 20 and 60 microns. Red cedar and Western ragweed pollen are on the low end, around 20 to 30 microns. Scot's pine and Carolina hemlock are between 55 and 80 microns. Cedar pollen is around 30 microns in diameter. Giant ragweed pollen grains are around 18 microns in diameter, and Noble Fir pollen is around 140 microns. See Figure 4.2 for representative types.

Fungal spores are in the form of spheres, ovals, spirals, elongated stellates (star-shaped), and clubs. They may be elongate, chained, or compact, and, generally, the surfaces are smooth. See Figure 4.3 for some shape differentiating features. They lack hairs, spicules (needles), and ridges, features common to pollen grains, which are more complicated in design.

Pollen grains tend to be spherical or elliptical with surface structures and/or pores, and the interior portions typically have a recognizable arrangement. They may be lobed with a smooth surface or spherical with spicules. Their interiors may be thick walled, undifferentiated or thin walled, multifaceted. Ragweed pollen is spherical with multiple spines, and pine pollen grains are lobed with a smooth surface.

Plant pollen is generally more complicated in design than are the spores. They tend to be spherical or elliptical with surface structures and/or pores, and the interior portions typically have a recognizable arrangement. They may be lobed with a smooth surface or spherical with spicules. Their interiors may be thick walled, undifferentiated or thin walled, multifaceted. Ragweed pollen has a spherical morphology with multiple spines. The pine pollen are lobed with a smooth surface.

Pollen densities range from 19 to 1,003 grains per microgram. Hickory pollen is moderate in size, weighing in on the low end of the scale. Giant Ragweed and nettle, even though small in size, are on the high end in density.

Spore-Producing Fungi and Bacteria

Both fungi and fungi-like bacteria produce allergenic spores. Although, the most commonly encountered spores in indoor air quality are mold spores, other fungal spores and bacterial spores can and frequently do contribute to the total airborne spore count.

Fungi

Fungi, numbering over 100,000 different species, are neither plant nor animal. Lacking in chlorophyll (plant-like) and motility (an animal characteristic), they belong to a kingdom of their own. The Fungi Kingdom consists of molds, yeasts, and mushrooms. Where the yeast are single-celled organisms, molds grow into long, tangled strands of cells that multiply, forming visible colonies of varying sizes, shapes,

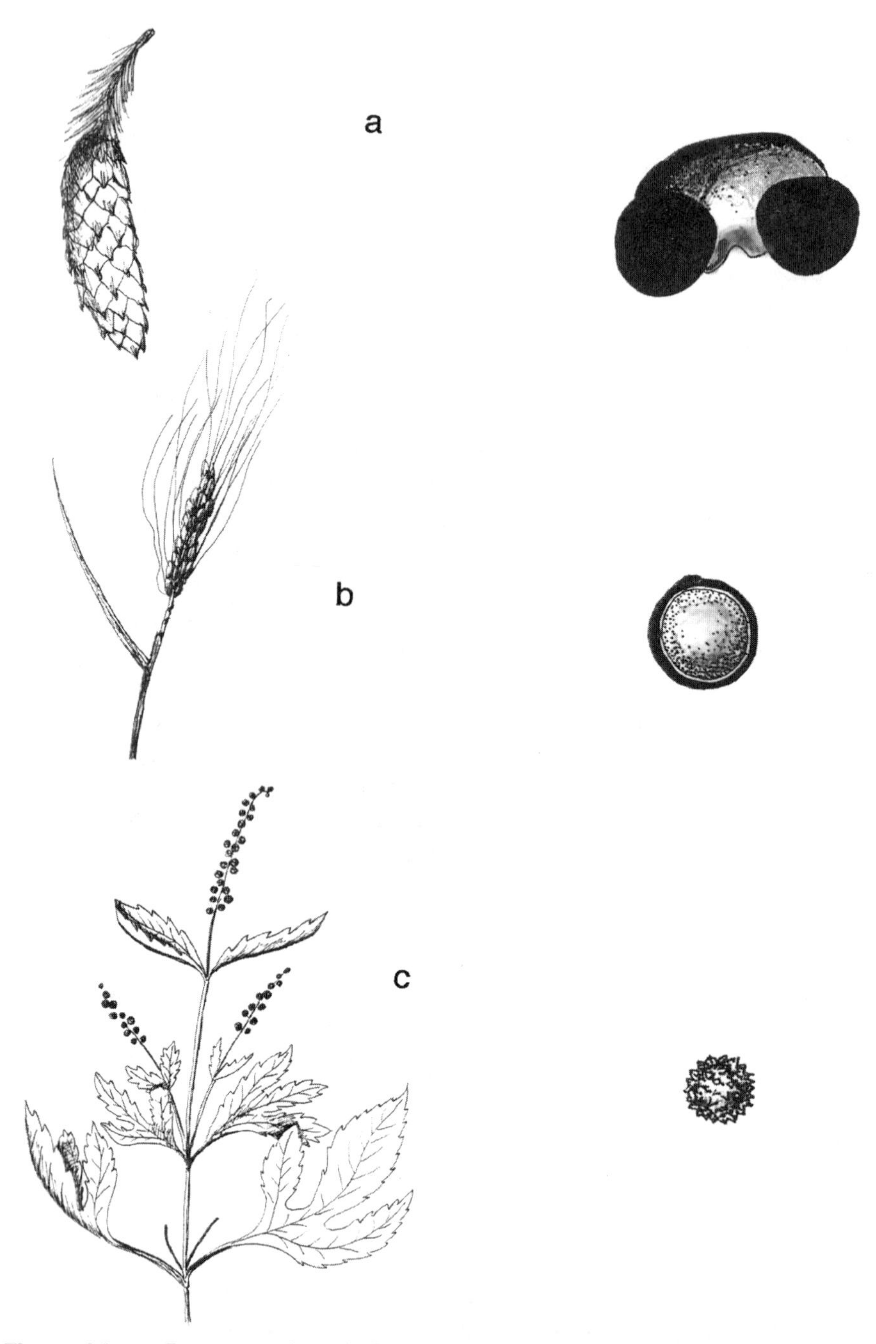

Figure 4.2 Representation of allergenic plants and pollen categorized into trees [e.g., cedar (a)], grasses [e.g., tall wheat (b)], and weeds [e.g., Giant Ragweed (c)]. The examples shown above are amongst the more allergenic within their category.

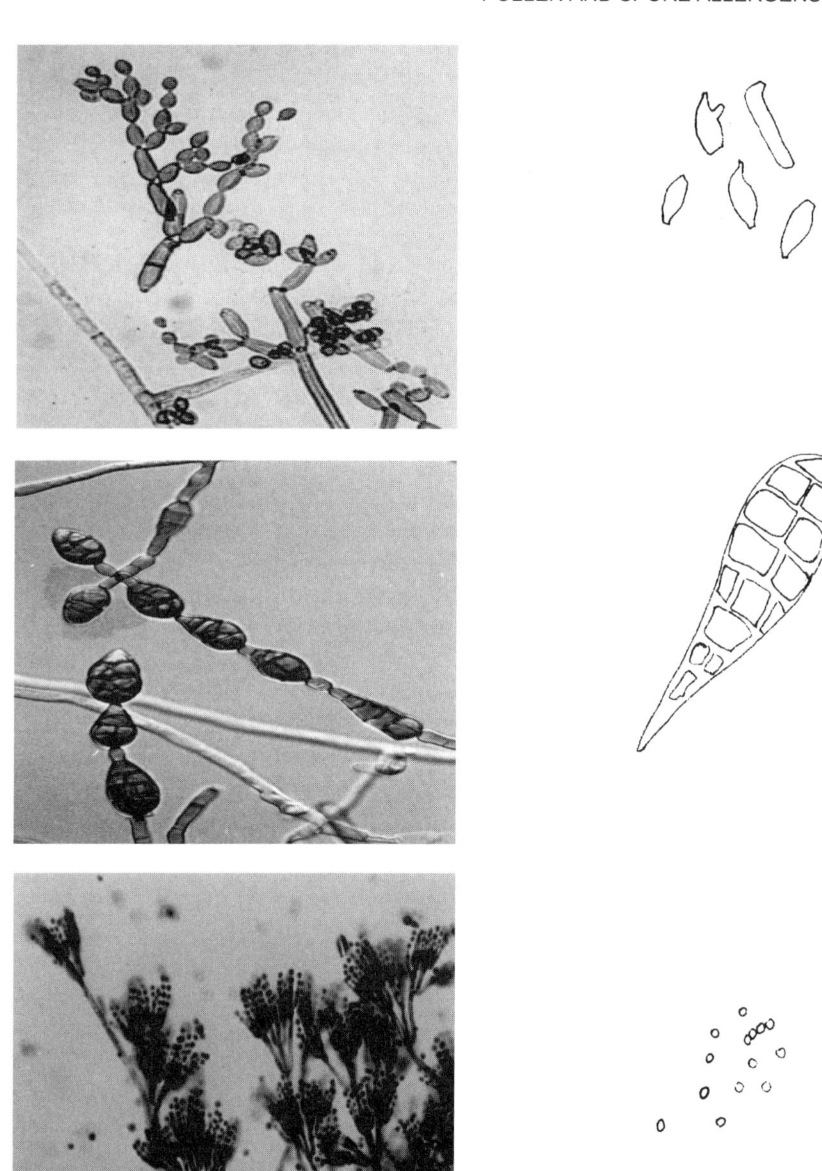

Figure 4.3 *Cladosporium* (top), *Alternaria* (middle), and *Penicillium* (bottom), are among the more commonly encountered mold spores in the outdoor air environment. The sketches on the right are relative size comparisons of the respective spore types. Photos contributed by Environmental Microbiology Laboratory, Inc. in Daly City, CA.

diameter (for pumpkin pollen). Grass pollen grains are usually around 20 to 40 and coloration. The more complex fungi are tightly compacted masses of mold-like forms (e.g., mushrooms).

Molds

Mold spores are the most commonly referred to fungal allergens. Their cell wall and protective spore surface is composed of polysaccharides (e.g., cellulose) and glucose units containing amino acids (e.g., chitin). The cellulose component is plant-like, and the chitin component is animal-like. It is the outer protective surface of the molds that is thought to be that which elicits an allergic reaction. Although spores are generally implicated to most allergy conditions, sections of the growth structures can be allergenic as well.

Mold reproduction involves the release of thousands of allergenic spores, each having the ability to reproduce long, thread-like hyphae that continue to branch and form mycelia. The mycelium, in turn, attaches to a nutrient substrate and grows. As long as the mycelium has nutrients and room to grow, a single mycelium may theoretically expand to a diameter of fifty feet. See Figure 4.4 for diagram of mold structures.

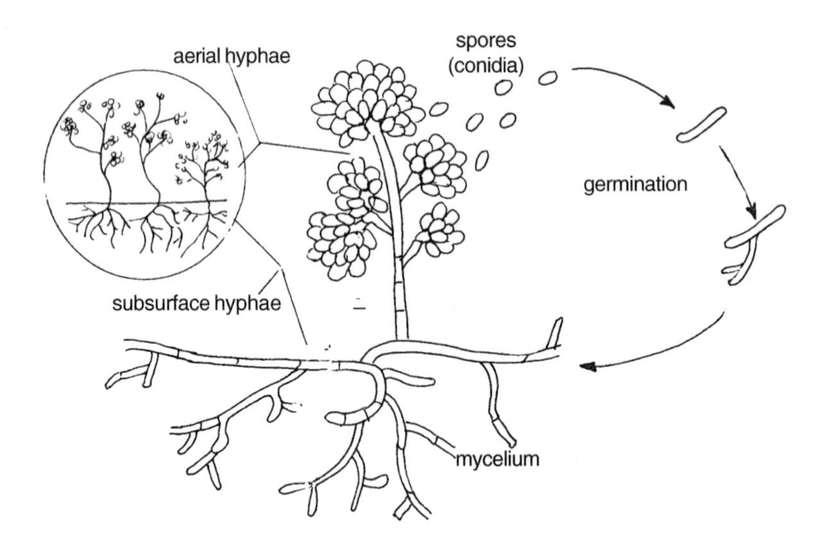

Figure 4.4 Typical Mold Structures

Specific mold genera are reputed to provoke allergy-like symptoms more consistently than others. This may be due to the challenge by shear numbers of a given species, or it may be due one species being able to illicit a stronger reaction than another. See Table 4.3. It is not clear as to which is the case.

Table 4.3 Mold Spores (in alphabetical order) Reported to Provoke Allergy Symptoms

Acremonium spp.[2]	*Mucor spp.*[4]
Alternaria spp.[2-5]	*Nigrospora spp.*[4]
Aspergillus spp.[2-5]	*Penicillium spp.*[2-5]
Aureobasidium spp.[4]	*Phoma spp.*[4]
Aureobasidium pullulans[2]	*Rhizopus spp.*[4]
Chaetomium spp.[4]	*Scopulariopsis*[4]
Cladosporium spp.[4]	*Stachybotrys spp.*[2]
Drechslera spp.[2 & 4]	rust molds and smuts[5]
Epicoccum spp.[4]	
Fusarium spp.[4]	
Helminthosporium spp.[5]	

Molds not commonly known to cause an allergic reaction may also contribute to the overall response of an individual's immune system to those molds that are reported to provoke allergy symptoms. Then, too, some authorities believe that individuals can develop an allergy to non-allergenic fungi or become sensitized to a fungal spore that is not commonly a problem for most people. Generally, however, allergenicity is genus specific. An allergic reaction to one mold type does not necessarily follow that the same will occur with another.

The most common airborne spore is *Cladosporium.* Beyond *Cladosporium*, there is some variation, based on geographic region and the time of the year. The consensus appears to be for *Alternaria* as the second largest contributor, and many include *Aspergillus* and *Penicillium.* Ironically, most of these molds are reputed causative allergenic agents for most mold-sensitive patients. Table 4.5 provides percents of total airborne mold spores reported by one source to represent the most common airborne allergenic molds. Most findings include many of the same genera with a slight variation on percent, based on regional differences.

A single colony is capable of dispersing millions of spores in one day. See Table 4.4. The spores are shot out of their capsule or dislodged from their stalk and carried by the wind to be spread far and wide. Spores (and pollen) travel, in extreme cases, as far as 1,500 miles,[6] and it is common to find them a hundred miles from their point of origin. More simply stated, their source does not necessarily have to be in the immediate vicinity.

Table 4.4 Number of Spore Discharges from One Source

Ganoderma applanatum	30 billion/day
Daldinia concentrica	100 million/day
Penicillium spp.	400 million/day

Excerpted from *Sampling and Identifying Allergenic Molds.*[7]

Table 4.5 Most Common Airborne Allergenic Molds from Nineteen Random Surveys

Genera	Prevalence (%)	Natural Habitats
Cladosporium	29.2	Worldwide: Soil, textiles, foodstuffs, and stored crops
		Region dependent: Woody plants (e.g., straw) and paints
Alternaria	14.0	Worldwide: Decaying plant matter, foodstuffs, soil, and textiles
Penicillium	8.8	Region dependent: Soil, decaying vegetation, foods, cereals, textiles, and paints
		Occasional occurrences: Composts, animal feces, paper/paper pulp, stored temperature foods, cheeses, and rye bread
Aspergillus	6.1	Region dependent: Soil, stored cereal products, soil, foodstuffs, dairy products, textiles, compost, and house dust
Fusarium	5.6	Worldwide: Soil and plants
Aureobasidium	4.7	Worldwide: Soil, decaying pears and oranges, paint, wood, and pape

Excerptec from Mould Allergy.[5]

Mushrooms

Mushrooms are filamentous fungi that typically form large structures, called fruiting bodies. These fruiting bodies are united to form what is commonly referred to as the mushroom cap. The reproductive mechanism is completed within this cap, and spores are discharged into the air, contributing to the total airborne fungal spores. Although it is unlikely that mushrooms will grow indoors without intent (e.g., cultivation of edible mushrooms), mushrooms have, on rare occasion, been reported growing in obscure, moist areas inside building structures. See Figure 4.5 for life cycle.

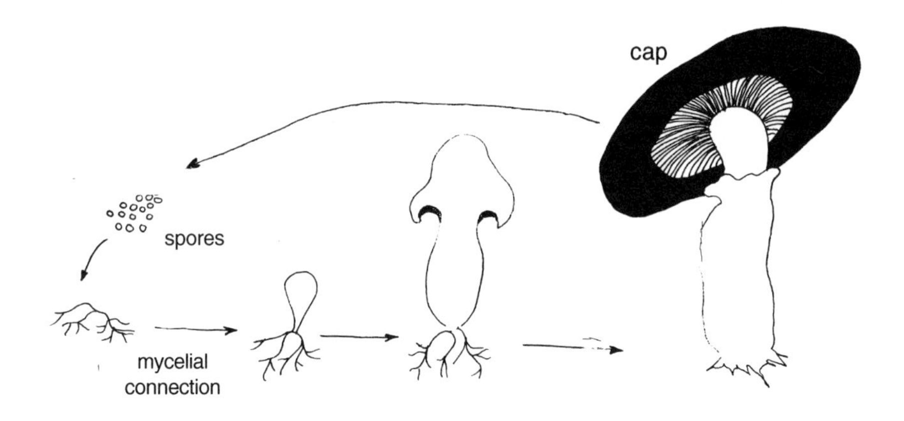

Figure 4.5 Life Cycle of Mushrooms

Rusts and Smuts

Single-celled rust and smut proliferate to form thick-walled, binucleate spores. With an excess of 20,000 species, "rust" fungi are so referenced due to an orange-red color imparted to diseased plants when the plants become infected. Heavily infected plants look like they are covered with iron oxide rust. Rusts do not grow indoors unless their host plants are present and infected.

"Smut" fungi have over 1,000 species. The term smut is assigned to this class of fungi, because the thick-walled spores impart a black, sooty appearance to plants. The levels of smut indoors are generally equal to or less than that of the outdoor air.[8] If the smut and rust spore levels equal the outdoor air, the fresh air is not adequately filtering the air.

Slime Molds

Slime molds are not true fungi, because they lack, for most of their lives, a cell wall. Laboratory reports refer to slime molds by their taxonomic fungal category-myxomycetes. The term slime mold refers to the swarming bodies of amoeboid cells during part of their life cycle. In this stage, many of the slime molds display brilliant colors and appear mucoid on a nutrient surface.

The slime molds have an interesting life cycle that includes a wet, amoeboid-like phase and a dry spore phase. See Figure 4.6. When conditions are favorable, they live primarily on decaying plant matter (e.g., leaf litter and logs) and bacteria-rich soils. Their food consists mainly of other microorganisms (e.g., bacteria and yeasts), and they ingest by phagocytosis. During the wet phase, they do not pose a problem.

During the dry phase, however, they form stalks that produce spores (or multiple spore-containing sporangia) that are subsequently released into the air. Slime mold spores can contribute to the total fungal spore count. It should be noted, however, that their spores may easily be mistaken for smut.

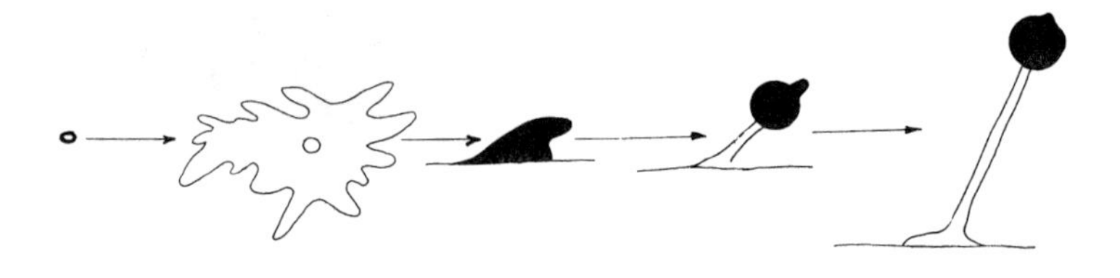

Figure 4.6 Life Cycle of Slime Molds

Bacteria[6]

Bacteria are single-celled organisms usually less than 1 micron in diameter, but they can be as large as 5 microns. The actinomycetes are filamentous bacteria that can produce structures which have the appearance of *Apergillus* and *Penicillium* mold spores and can contribute to the total allergenic spore count. Their spores are also allergens.

Actinomycetes are spherical or oval in shape and range in diameter from 0.8 to 3 microns, similar to that of *Aspergillus/Penicillium*. However, their nutrient requirements are complex. They grow best in rich organic material and tolerate extremes in temperature. Thus, they do not grow in conditions similar to those found in most office building. Differentiating the mold and bacteria spores can only be accomplished by culturing viable spores.

Bacilli are rod-shaped, spore-forming bacteria. They are generally associated with food spoilage and are not likely to be airborne.

Indoor Source Information

Although indoor pollen grain exposures are generally less than outdoor exposures, the reduced pollen count indoors may still contribute to the total allergen loading to which an individual is exposed. There are, also, exceptions to the rule in that indoor pollen counts, on rare occasion, may be greater than outdoor counts. In the latter case, pollen may have entered a building during the pollen season when the windows and doors were open and once inside the building, the pollen enters into a recycling mode within a poorly filtered air handling system. The investigator should not be blinded to all possibilities.

Fungal spores, also, enter the indoor air from outside, and the total spore count is generally less indoors than it is outside. With mold spores, the total indoor mold spore count is typically 10 to 50 percent less than the outside air. Yet, unlike the pollen grains, molds and occasionally other fungi can grow indoors. Not only does their growth contribute to the total count, but some types of molds pose a greater health concern than others. For this reason, an effort to identify fungal types is necessary to characterize their impact.

Once again, there are exceptions to the rule in that indoor mold spores may on occasion be greater indoors than outside. When this occurs at low levels (e.g., less than 200 counts/m^3 outside), it may be the result of outside conditions that have minimized the outside mold spore count (e.g., immediately after a rain, which tends to settle particles and mold spores out of the air), or it may be the result of normally low outside levels with amplification, or growth, of molds indoors. Normal conditions will come with experience in air sampling within a given region. For instance, outside air in St. Louis, Missouri is normally in excess of 2,000 counts/m^3 (as high as 62,000 counts/m^3) whereas outside air in Las Vegas, Nevada is normally less than 100 counts/m^3 (rarely higher than 2,000 counts/m^3). A clear case of amplification indoors would be an outside level of 11,000 count/m^3 and an indoor exposure of 44,000 counts/m^3.

Fungal species have different growth requirements, habitats frequented, health effects, and levels of concern. Yet, to capture and count all allergenic fungal spores, the investigator must settle for a more generalized characterization of fungal spores, and proceed to Chapter 5 for identification of genus and, in some cases, species. See Table 4.4 for information regarding fungal identification generally within the scope of the sampling methodology presented within this chapter.

SAMPLING STRATEGY

The investigator should determine the purpose for air sampling, and the purpose should assist the investigator to identify the area(s) to be sampled. They may be based on identification of one, or a combination of the following: (1) perceived worse case scenario(s); (2) representative of area(s) frequented; and (3) areas of special concern (e.g., infant nursery). These sample areas should be compared to at least one outside air sample and, if possible, one non-complaint area such as may occur in an office building. Once the site or sites have been selected, sample duration should be considered.

The occupants' health and exposure duration are factors to consider when deciding the type of air sampling to be performed. The exposure considerations for a healthy adult at work in an office building will be different from that of an elderly patient confined to a hospital bed.

Activities should, also, be taken into consideration. Some activities may impact the exposure levels more than others. For instance, maintenance removing ceiling tiles in an office building may result in a two to three fold increase in the spore counts. Custodial vacuuming and dusting may result in increased counts. Aggressive agitation of bedding, textiles, and clothing may result in increased counts. A humidifier may result in decreased counts. Sometimes these peak exposure activities go unnoticed without more extensive samples per site throughout the day or a time-discrete sampler.

Based on each scenario, the investigator must decide on the appropriate sampling method. These methods are typically short-term, snapshot samples of several sites or short-term, time-discrete samples over a period of time.

SAMPLING AND ANALYTICAL METHODOLOGIES

In indoor air quality sampling, the sampling methodologies of choice are the slit-to-cover-slip sample cassettes (e.g., Air-O-Cell) and slit-to-slide samplers (e.g., Allergenco™ Spore Trap). They all perform on a similar principle but vary in up-front cost, on-going expenses, and ease of handling large sample numbers.

Try not to compare sample results taken by two different approaches. Taking two samples using the same sampler at the same time and same approximate location may result in differences. Comparing samplers is a job for researchers. Choose one sampler, and stay with it.

Slit-to-Cover Slip Sample Cassettes

The slit-to-cover slip sampling methodology is often referred to as Air-O-Cell sampling. The Air-O-Cell is a cassette with a slit opening through which air passes and particles adhere to the surface of a sticky substance (e.g., triacetin) on the surface of a cover slip. The air is drawn through the cassette by means of an air sampling pump.

During sampling, the protective tape on either side of the Air-O-Cell is removed. The cassette is connected to a calibrated air sampling pump, air is sampled for an abbreviated time period, and the cassette is re-sealed and sent to a laboratory for analysis. A summary of the method follows:

- Equipment: air sampling pump and timer
- Collection medium: Air-O-Cell cassette

- Flow rate: 15 liters/minute
- Recommended sample duration: 1 to 10 minutes, based on anticipated loading

Anticipated loading is based on conditions and activities. Where excessive loading occurs on the slide, enumeration becomes difficult if not impossible. In the latter case, samples may be significantly underestimated and difficult to identify.

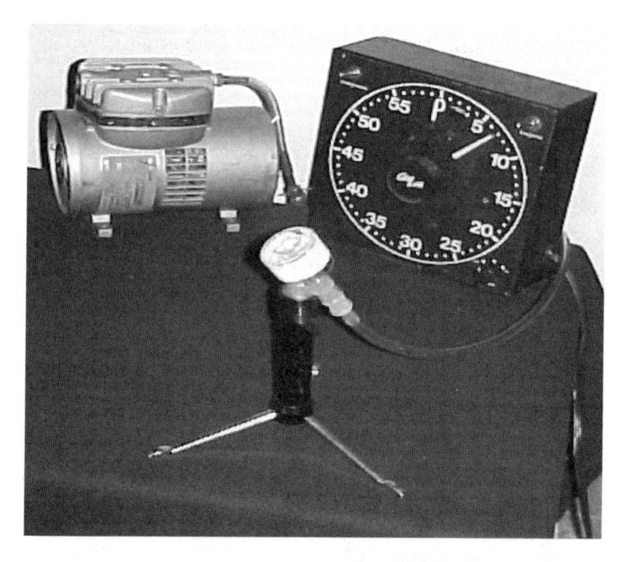

Figure 4.7 Air-O-Cell Sampling with Air Sampling Pump and Timer

In clean office environments and outside where there is very little dust anticipated, sampling should be performed for 10 minutes. In dusty areas and/or areas where there is considerable renovation, a 1-minute sample should be considered. Indoor air environments where there is moderate dust or where considerable levels of mold spores (e.g., greater than 500 spores) are anticipated, the sampling duration should be reduced accordingly (e.g., 6 to 8 minutes). Experience will be the investigator's best guide.

Slit-to-Slide Samplers

The slit-to-slide sampler operates by impacting particles onto a treated microscope slide. An internal pump draws air through the slit at a flow rate of 15 liters per minute. Sampling duration may be from 1 to 10 minutes, and the samplers can be programmed to collect a different sample at designated sample intervals. For instance, the Allergenco™ Spore Trap can be programmed to collect 24 discrete samples, once an hour for a total of 24 samples per slide.

With the ability to program the sampler, the investigator may find the slit-to-slide sampler easier to manage than the cassettes. One location can be assessed over an extended period and trends documented for time verses concentration.

- Equipment: Burkard™ 7-Hour Spore Trap or Allergenco™ Spore Trap
- Collection medium: treated microscope slide
- Flow rate: 10- to 15 liters/minute
- Recommended sample duration: 5 to 10 minutes, based on anticipated loading

Figure 4.8 Allergenco™ Spore Trap with Built-in Timer, Programmable for up to 24 Discrete Samples on One Slide

Analytical Methods

Samples are received, stained, covered (e.g., cover slip on the microscope slide), and looked at under an optical microscope. Counts are generally performed with a 40x objective, and identification is performed using the 100x oil immersion.

The entire impacted surface area should be counted and results given in terms of fungal spore characterization (e.g., myxomycetes) as well as spore and pollen counts per cubic meter of air. Characterization identifies fungal spores by categories (e.g., rusts) and mold spores by groups (e.g., *Aspergillus/Penicillium*), and some of the more morphologically unique molds by genus (e.g., *Stachybotrys*).

Commercial Laboratories

In the recent years, commercial laboratories have been popping up like mushrooms after a heavy rain. They came, they saw, and they conquered. Although there are attempts underway by universities (e.g., Harvard School of Public Health) and nationally recognized organizations (e.g., the American Industrial Hygiene Association) to certify analysts and laboratories, many analysts are inexperienced, and there is no quality control watchdog.

On the other hand, there are some very competent, experienced laboratories that charge a little extra, are not local, or have longer turn around times. The trade-off may be worth it. It can be disconcerting to compare two separate laboratory results of the same samples only to find one laboratory state the counts are excessive while the other fails to detect any levels. If there is a divergence from any of the basic laboratory approaches mentioned above or the results appear inconsistent, seek a second opinion.

Helpful Hints

A dilemma that many investigators ponder is what equipment to purchase. Should the investigator choose to perform time-discrete sampling, the number of sample sites is limited by the number of impactors available. Multiple sampling of several site may be performed by any of the previously mentioned methods, and time-discrete sampling may be performed by the environmental unit Burkard™ Spore Trap and the Allergenco™ Air Sampler.

The handling of the treated slides that are used in the impactor must be done with caution. Once treated, the slides must be kept clean and should not be touching other surfaces, including other microscope slides. The slide may be maintained and transported in a box or plastic container specifically made for this purpose.

This is generally not a problem with the Air-O-Cells. The Air-O-Cells are sealed prior to sampling and should be resealed upon sample completion.

Although the initial cost of the air sampling pump for use with the Air-O-Cells is less than the slit-to-slide impactors, the cost of the Air-O-Cells cassettes can outweigh the initial cost of the impactors.

Due to its limitations, the Rotorod™ has not been discussed herein. Used in the past for community allergy alert reporting, the Rotorod™ has a rotating rod with a built-in 24-hour interval timer. It is easy to use, and the results are read in terms of counts/m^3. Its limitation, however, is in the size of particles impacted onto the rod. There is a considerable drop off of smaller particles, particularly the spores that are most commonly found to be problematic in indoor air quality (e.g., *Aspergillus* and *Penicillium* molds).

When not able to program the sample time, use a good timer—preferably one that counts down in seconds. Some timers count down in minutes, and it is difficult to anticipate stop sampling time. The author prefers a timer with a built-in turn off switch (e.g., darkroom clock). It is easy to become distracted while waiting for the sample to be collected.

Keep notes of conditions at each sample location. These should include exact location within the room, air movement (e.g., air supply not on while sampling), distance from air supply vents, occupancy (e.g., nighttime no occupancy), and activities (e.g., busy area). Many investigators also log temperature, relative humidity, and carbon dioxide levels.

INTERPRETATION OF RESULTS

Although there are no indoor air quality standards for interpreting pollen and mold spore counts, the National Allergy Bureau has set some guidelines based on ecological measurements for outdoor air. As health effects are dependent on individual susceptibility, the relative exposure index is not based on health effects. They are relative numbers and limits. Yet, an investigator may excerpt in part-or-parcel usable information from these tables. See Table 4.6.

Table 4.6 National Allergy Bureau Guideline for Relative Exposures to Outdoor Air Pollen and Spores (counts/m^3)[8]

Allergen	Very Low	Low	Medium	High	Very High
Molds	<500	500-1,000	1,000-5,000	5,000-10,000	>20,000
Pollen	1-50	50-100	100-500	500-1,000	>1,000

The National Allergy Bureau is a section of the American Academy of Allergy, Asthma, and Immunology (AAAAI) Aeroallergen Network that is responsible for reporting current pollen and mold spore levels to the media. The Network is a group of pollen and spore counting stations staffed by AAAAI members who volunteer to donate their time and expertise to providing the most accurate and reliable pollen and mold counts from over 65 counting stations throughout the United States and Canada. They use the 7-day long-term Burkard™ Spore Trap in the performance of their sampling.[9]

Beginning in 1992, the National Allergy Bureau has compiled records reported by each of the stations. These are broadcast to the media, and they are posted on the Bureau's website (www.aaaai.org/nab). These records and additional allergy information can be accessed by the public.

As for assessing indoor air quality, sample results may require additional considerations. The indoor counts should be compared to outdoor counts and, if possible, to indoor noncomplaint areas, and the types of fungal spores found indoors verses those found outdoors may be compared when assessing the potential source of mold spores as being indoors.

The nonviable spores are more difficult to identify, but they can be characterized. Identification is not as complete with total pollen and spore air monitoring methodologies, but characterization and some identification may be useful in determining amplification indoors. See Tables 4.7 and 4.8 for generalized categories used in laboratory reports and for additional information regarding occurrence.

SUMMARY

There are no standards for interpreting pollen/spore counts and their effect on human health. The investigator can assess total relative pollen and spore counts and, to a limited extent, compare types to determine amplification indoors. For a more thorough identification and comparison of types in determining indoor amplification of molds, the investigators should also perform viable mold air sampling in tandem with the viable/nonviable methods in this chapter. See Chapter 5.

REFERENCES

1. Aeroallergen Network of the American Academy of Allergy, Asthma and Immunology (AAAAI). U.S. Pollen Calender. AAAAI, Milwaukee, Wisconsin, July 1993.
2. Smith, E. Grant. *Sampling and Identifying Allergenic Pollens and Molds: An Illustrated Identification Manual for Air Samplers.* Blewstone Press, San Antonio, Texas, 1990. p. 43.
3. ACGIH. Bioaerosols: Airborne Viable Microorganisms in Office Environments—Sampling Protocol and Analytical Procedures. *Applied Industrial Hygiene*, April 1986. p. R-22.
4. Cole, Garry T. and Harvey C. Hock. *The Fungal Spore and Disease Initiation in Plants and Animals.* Plenum Press, New York, New York, 1991, p. 383.
5. Al-Doory, Yousef and Joanne F. Domson. *Mould Allergy.* Lea & Febiger, Philadelphia, Pennsylvania, 1984. pp. 36-37.
6. Smith, E. Grant. *Sampling and Identifying Allergenic Pollens and Molds: An Illustrated Identification Manual for Air Samplers.* Blewstone Press, San Antonio, Texas, 1990. p. 16.
7. Ibid. p. 42.
8. Pinnas, Jacob L. Parameters for Pollen/Spore Charts. [Letter] National Allergy Bureau, Tucson, Arizona, November 28, 1994.

9. Bleimehl, Linda. Issues concerning the National Allergy Bureau. [Oral communication] AAAAI, Milwaukee, Wisconsin, January 1996.
10. Gallup, Janet and Miriam Valesco, Dr. P.H. Characteristics of Some Commonly Encountered Fungal Genera. Environmental Microbiology Laboratory, Inc., Daly City, Calif. (1999).

Table 4.7 Characteristics of Molds

Genus	Where Found in Nature	Conditions to Grow Indoors
Acremonium	soil, dead organic debris, hay, food stuffs	very wet conditions
Alternaria	soil, dead organic debris, food stuffs, textiles (some plant pathogens)	>85% moisture
Arthrinium	soil, decomposing plant material	cellulose rare
Aspergillus	soil, decaying plant debris, compost piles, stored grain	>70% moisture
Aureobasidium	soil, forest soils, fresh water, aerial portion of plants, fruit,	widespread where moisture, marine estuary sediments, wood accumulates, especially in bathrooms and kitchens, on shower curtains, tile grout, window sills, textiles, liquid waste
Beauveria	soil, plant debris, dung, insect parasites	
Botrytis	soil, stored and transported fruit and vegetables, plant pathogen, saprophyte on flowers, leaves, stems, and fruit, leaf rot on grapes, strawberries, lettuce, cabbage, onions	>93% moisture
Ceratocystis/	commercial lumber, tree and plant pathogen	grows on lumber wood framing in residences
Cercospora	parasite of higher plants, causing leaf spot	
Chaetomium	soil, seeds, cellulose substrates, dung, woody and straw materials	cellulose, damp sheetrock
Cladosporium	soil of many different types, plant litter, plant pathogen, leaf surfaces, old or decayed plants	textiles, wood, moist window sills grows @ 0°C, >85% moisture
Curvularia	plant debris, soil, facultative plant pathogens of tropical and subtropical plants	variety of substrates
Drechslera, Bipolaris, and *Exserohilum*	plant debris, soil, plant pathogens (particularly grasses)	variety of substrates
Epicoccum	plant debris, soil, secondary invader of damaged plants	paper, textiles, and insects; >86% moisture
Fusarium	soil, saprophytic or parasitic on plants, plant pathogens	variety of substrates; >86% moisture
Memnoniella	plant litter, soil, many types of plants and trees	cellulose, variety of substrates
Mucor	organic matter, dung, soil	variety of substrates, including leftover food, soft fruits, and juices; >90% moisture

Table 4.7 Characteristics of Molds (continued)

Genus	Where Found in Nature	Conditions to Grow Indoors
Myrothecium	grasses, plants, and soil; decaying fruiting bodies of *Russula* mushrooms	
Nigrospora	decaying plant material and soil	
Paecilomyces	soil and decaying plant material, composting processes, legumes, cottonseeds, some species parasitize insects	jute fibers, paper, PVC, timber (oak wood), optical lenses, leather, photographic paper, cigar tobacco, harvested grapes, bottled fruit, and fruit juice undergoing pasteurization; >80% moisture
Periconia	soil, blackened and dead herbaceous stems and leaf spots, grasses, rushes, sedges	
Phoma	plant material, soil, fruit parasite	walls, ceiling tiles, reverse side of linoleum; cement, paint, paper, wood, wool, and foods such as rice and butter; spores not readily disseminated by air currents
Pithomyces	dead leaves, soil, grasses	paper
Rhinocladiella	soil, herbaceous substrates, and decaying wood	variety of substrates, found around wine cellars on brickwork and adjacent timber
Rhizopus	forest and cultivated soils, decaying fruits and vegetables, animal dung, and compost; a parasitic plant pathogen on cotton potatoes, and various fruits	variety of substrates, spoiling food; >93% moisture
Sporobolomyces	tree leaves, soil, rotting fruit, other plant materials, associated with lesions caused by other plant parasites	variety of substrates very wet conditions
Stachybotrys	soil, decaying plant substrates, decomposing cellulose (hay, straw), leaf litter, and seeds	cellulose (e.g., wallboard, jute, wicker, straw baskets, other paper materials); >94% moisture
Stemphylium	soil, wood, decaying vegetation, some species plant pathogens	
Torula	soil, dead herbaceous stems, wood, grasses, sugar beet root, ground nuts, and oats surface of unglazed ceramics, and cellulose	cellulose

Table 4.7 Characteristics of Molds (continued)

Genus	Where Found in Nature	Conditions to Grow Indoors
Trichoderma	soil, decaying wood, grains, citrus fruit, tomatoes, sweet potatoes, paper, textiles, damp wood	paper, tapestry, wood, in kitchens on the outer
Ulocladium	soil, dung, paint, grasses, fibers, wood, decaying plants, paper, and textiles	gypsum board, paper, paint, tapestries, jute, and other straw materials; high water requirement, relatively dry surfaces
Wellemia sebi	soil, food stuffs, hay, textiles, salted fish	>69% moisture; wood in crawl spaces, mattress dust; may colonize on human skin cells

Excerpted from "Characteristics of Some Commonly Encountered Fungal Genera."[10]

Table 4.8 Categories of Fungal Spores

Types of Fungi	Where Found in Nature	Conditions to Grow Indoors
Ascospores: *morels, truffles, cup fungi, ergot, and many micro-fungi*		
	saprophytes and plant pathogens	damp substrates
Basodospores: *mushrooms, puffballs, shelf fungi, bracket fungi, earth stars, stinkhorns, and other complex fungi*		
	saprophytes and plant pathogens gardens, forests, woodlands rot, damage structural wood of buildings	"dry rot" and wood, poisonous
Coelomycetes: *asexual fungi that form conidia in a cavity or mat-like cushion of hyphae*		
	Saprophytic or parasitic on higher plants, other fungi Lichens, vertebrates	ceiling tiles, linoleum; spores not readily disseminated by air currents
Myxomycetes: *slime molds*		
	Decaying logs, stumps and dead leaves, particularly in forested regions	
Rusts	grasses, flowers, trees, and other	indoor plants
Smuts	cereal crops, grasses, weeds, other fungi, and other flowering plants	indoor plants

Excerpted from "Characteristics of Some Commonly Encountered Fungal Genera."[10]

Chapter 5

VIABLE MICROBIAL ALLERGENS

Microbial allergens are microscopic organisms that may cause allergy symptoms. Mold spores are the most commonly recognized microbial allergens, reported to affect over 20 percent of the adult population. Other less commonly recognized microbial allergens are spore-forming bacteria.

Microbial organisms may be culturable (i.e., grown under laboratory conditions), nonculturable, or dead, and they may illicit the same effect, irrespective of viability. On the other hand, not all microbes are allergens. Some are human pathogens (e.g., tuberculosis). Many affect our lifestyles (e.g., mildew in the home) and crops (e.g., crop mold). Others have no apparent, direct impact on our lives. Yet, they may all be airborne.

Airborne microbial allergens are indoors and outdoors. Generally, the outdoor levels are greater than the indoor levels. Even when the outdoor levels exceed the indoor levels, identification of the indoor and outdoor microbial allergens by genus and species is a means for determining amplification indoors. This is done by culturing viable microbes. A viable organism is one that is capable of growing and completing a life cycle. Thus, microbes can only be reliably differentiated when their viability is retained. If they are dead, cannot grow on culture media, or cannot form spores when grown on culture media, microbial allergens cannot be identified.

Although the health effects of viable allergens and nonviable allergens are the same, sample collection and interpretation methods are distinctly different. Airborne viable microbial air sampling is more involved than nonviable sampling. Viable air sampling is the only way to definitively identify mold spores by genus and species, but other fungal spores (e.g., smuts and rust) are rarely cultured. In the latter, the most efficient way to sample is for viable/nonviable spores.

The information obtained through viable sampling contributes significantly toward the overall picture and interpretation of exposures to mold spores. Mold spores are a major contributor to indoor air quality allergens.

OCCURRENCE OF ALLERGENIC MICROBES

Viable allergens are herein discussed to lend an understanding to the reader as to the potential sources and growth requirements needed for enhancement and amplification of allergenic microbes. The principal allergen in each case is thought to be airborne spores with minimal concern for the associated growth structures. The two major categories are fungi and thermophilic actinomycetes.

Those microbes which have been excluded from this allergenic microbes list are either pathogenic or have not been reported as probable allergens. The pathogenic microbes are discussed separately in Chapter 6.

Fungi

Molds and other fungal allergens were discussed in the preceding chapter under the section "Spore-Producing Fungi and Bacteria." Yet, the intent was to discuss basic characteristics, not differences between the specific fungi. As they can be more readily identified, we discuss herein details regarding specific fungi, their habitats, and unique characteristics.

Molds

Contrasted with the hardiest of microbes, molds not only can stay alive indefinitely on inanimate objects, or fomites, but they can grow into and destroy wood, cloth, fabrics, leather, twine, electrical insulation, and many other commercial products. They destroy lenses of microscopes, binoculars, and cameras. In localities where humidity is high, fungi do great harm to wood structures, telephone poles, railroad ties, and fence posts. Most of these problems are reduced by means of artificial preservatives. Sometimes the trouble starts in forests where fungi invade the heartwood and cause wood rot before the timber has had a chance to be cut down. The humid Amazon rain forest is one such example.

Thousands of products are treated to prevent decay, yet there are some types of fungi that thrive on preservative-treated wood. One such example is creosote-treated railroad ties! Other fungi-specific nutrients are vinyl wall covering adhesives, gypsum board, cellulose-based ceiling tiles, dirt retained within carpeting, and surface paints. Some feed on plywood. Others consume the glue used to laminate wood that is used in airplanes, furniture, and cars and will cause the layers to separate. Books and leather shoes are readily consumed by the microbes that are visible as mildew. Aircraft electrical systems, operating in tropical climates, require protection against insulation-consuming molds. Immune-suppressed individuals (e.g., AIDS patients) can also be host to nonpathogenic, invading molds.

Exterior molds grow on decayed organic material, corn, wheat, barley, soybeans, cottonseed, flax, and sun-dried fruits. They consume other plants, vegetable matter, and decayed organic material (e.g., dead animals).

Pathogenic molds parasitize and obtain nutrients from a host. The host may be plant or animal, and these molds vary slightly from the nonpathogenic molds both in environmental and sampling requirements. Thus, they are discussed in greater detail in Chapter 6.

Most molds require high moisture content. Whereas most require moisture content in excess of 80 percent, some do quite well at levels as low as 60 percent. The latter are referred to as zeophylic (dry-loving) fungi. Whereas *Stachybotrys* requires considerable moisture, *Aspergillus versicolor* does well on slightly moist gypsum board. See Table 5.1 for water requirements of some of the more common microorganisms, and see Table 5.2 for moisture requirements of the more common fungi, identified by moisture preferences.

Table 5.1 Moisture Requirements of Common Microorganisms

Microorganism	Water Activity (% relative moisture)
Aspergillus halophilicus and *Aspergillus restictus*	0.65-0.70
Aspergillus glaucus and *Sporendonema sebi*	0.70-0.75
Aspergillus chevalieri, Aspergillus candidus, Aspergillus ochraceus, Aspergillus versicolor, and *Aspergillus nidulans*	0.75-0.80
Aspergillus flavus, Aspergillus versicolor, Penicillium citreoviride, and *Penicillium citrinum*	0.80-0.85
Aspergillus oryzae, Aspergillus fumigatus, Aspergillus niger, Penicillium notatum, Penicillium islandicum, and *Penicillium urticae*	0.85-0.90
Yeasts, bacteria, and many molds	0.95-1.00

Excerpted from *Microbiological Ecology of Foods.*[1]

Temperature preferences are variable as well. Although most molds do well at room temperature, some flourish at freezing temperature (e.g., refrigeration), and some thrive at temperatures in excess of 100°F (e.g., hot tubes). Although most spores favor moderate temperatures, the investigator should be aware of the potential for growth and amplification of molds in just about any temperature setting. The ranges are presented in Table 5.3.

Fungi tend to grow more during months when the humidity and temperature are elevated. In some regions, the peak mold spore season is in the spring, followed by the summer months. Other areas of the country experience peak periods in the fall. The winter months typically provide the least accommodating conditions for fungal growth. Although the daily and monthly variability is based entirely on humidity and temperature, growth will increase or decrease at certain hours of the day or night regardless of the outdoor climate.

Many, not all, fungi have peak growth times which are genus, sometimes species, dependent. Some peak in the late of night (e.g., *Cladosporium* and *Epicoccum*). Others peak in the early morning. Some peak in the late afternoon (e.g., *Alternaria* and *Penicillium*). A few peak, irrespective of time frame, immediately

Figure 5.1 Photomicrographs of allergenic molds with their growth structures. They are: *Penicillium* spp. (top left), *Scopulariopsis* spp. (top right), *Verticillium* (middle left), *Alternaria* (middle right), and *Aspergillus niger* (bottom left). Mold structures and spores stained, photos taken under 1000x oil immersion. Contribution by Environmental Microbiology Laboratory, Inc. in Daly City, California.

Table 5.2 Moisture Requirements of Common Fungi

Water Requirement	Common Indoor Fungi	Typical Sites
Hydrophilic fungi (>90% minimum)	*Fusarium, Rhizopus, Stachybotrys*	Wet wallboard, water reservoirs for humidifiers, and drip pans
Mesophilic fungi (80 to 90% minimum)	*Alternaria, Epicoccum, Ulocladium, Cladosporium, Aspergillus versicolor,*	Damp wallboard and fabrics
Xerotolerant (<80% minimum)	*Eurotium (Aspergillus Glaucus group),* some *Penicillium* species, and *Aspergillus restrictus*	Relatively dry materials (e.g., house dust where high relative humidity)
Xerophilic fungi (<80% preferred)	*Aspergillus restrictus*	Very dry and high sugar foods and building materials

Excerpted from *Bioaerosols: Assessment and Control.*[2]

after a heavy rainfall.[3] Studies vary on their opinion as to these times, yet they all agree that peak periods do exist.

Indoor air environments may vary in humidity content of the air along with the outdoor air environment, and peak mold seasons may correlate to the indoor environment, particularly where humidity controls are not maintained. High humidity (greater than 60% relative humidity) is thought to be the leading cause of fungal amplification within buildings.

Table 5.3 Temperature Relations of Common Mold Species

Temperature to kill most "mold spores" (within 30 minutes)	140-145°F
Maximum growth temperatures	86-104°F
Optimum growth temperature	72-90°F
Minimum growth temperature	41-50°F

Excerpted from *Manual of Medical Mycology.*[4]

Other means of mold spore amplification include, but are not limited to: (1) settled water sources (e.g., air handling system drip pans); (2) damp building materials (e.g., wet ceiling tiles); (3) air movement from a hot, humid crawl space into an occupied office area; (4) disturbances of settled dust (e.g., dry dusting); and (5) poor vacuum cleaner filtration. It should also be noted that mushrooms have been identified in buildings where water-damaged carpeting has been left uncorrected and other areas where there is an accumulation of water (e.g., behind leaking washing machines).

Molds grow in wide ranges of pH. Although a pH of 5 to 6 is favored by most, some molds are proliferate between pH 2.2 and 9.6. Rare forms have been found consuming nutrient impurities found in bottles of sulfuric acid.

As fungi typically require oxygen, they tend to grow in oxygen rich environments (e.g., air handling systems). Some, however, grow quite well in enclosed areas where the oxygen may be minimal (e.g., between vinyl wall coverings and the wall). Yet, they all do require some oxygen. There are no anaerobic fungi.

Yeasts

Yeasts, one-celled fungi, are usually spherical, oval, tube-shaped, or cylindrical in shape. They usually do not form filamentous hyphae or mycelium. A population of yeast cells remains a collection of single cells or budding structures. They can be differentiated from bacteria only in their size and internal morphology (with an obvious presence of internal cell structures). Some of the yeasts reproduce sexually. The sexual reproductive process forms an ascospore that is resistant to many environmental conditions. They tend to grow on nutrient agar intended for molds, and they are generally reported along with the molds on a cultured sample.

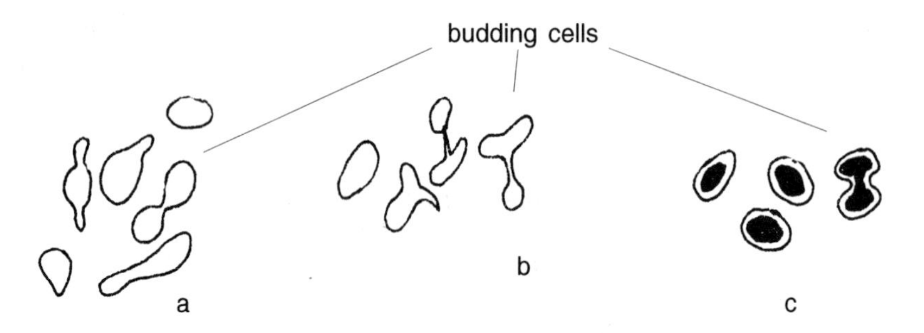

Figure 5.2 Yeasts *Rhodotorula* (a), *Sporobolomyces* (b), and capsulated *Cryptococcus neoformans* (c).

Yeasts usually flourish in habitats where sugars are present (e.g., fruits, flowers, and the bark of trees). The most important ones are the baker's and beer brewer's yeasts. These have been selected and manipulated by man. They do not serve as good representatives of the classification. They are atypical.

Although not common, yeasts have been reported growing indoors on wet, rotting wood and other high moisture content surfaces. When this occurs, the indoor yeast levels may exceed the outdoor levels. This is, however, rare. Those that are routinely found in indoor air quality investigations are *Rhodotorula* (shiny pink colonies on malt extract agar) and *Sporobolomyces* (shiny salmon-pink or red colonies on malt extract agar). *Cryptococcus neoformans* is a pathogenic yeast that produces a thick protective capsule.[5]

Bacteria[5]

Bacteria are single-celled organisms ranging in size from 0.8 to 5 microns in diameter. Their surface structure is complex, and they are limited in form. They may appear as spheres (e.g., cocci), straight rods (e.g., bacilli), or spiral rods (e.g., spirochetes), or branched filaments (referred to as actinomycetes).

Some bacteria produce endospores, or internal spores, which are resistant to environmental stresses. Endospores may be allergenic, and they may survive harsh conditions for extended periods. During this dormant period, endospores remain viable and allergenic. They can remain dormant for years. The most commonly known endospore-producing bacterium is the genus *Bacillus*, many species of which are also pathogenic. Actinomycetes normally produce spores that are readily released into the environment without the need for environmental stressors.

Both *Bacillus* and actinomycetes bacteria can be allergenic, and they require different nutrient agar than that required by molds. They require special media and the actinomycetes require elevated incubation temperatures. They are not likely to show up on mold culture plates.

Bacillus

The genus *Bacillus* is a gram-positive, rod-shaped bacterium that is known to form endospores under stress conditions. *Bacillus* endospores have been implicated as a cause of hypersensitivity pneumonitis, and the endospores have been found indoors as well as outdoors.

Bacillus bacteria thrive on dead or decaying organic material. The spores are normally found in soil, dust, and water. They can also be found in dry desert sands, hot springs, artic soils and water, pasteurized milk, stored vegetables, various foodstuffs, and feces.

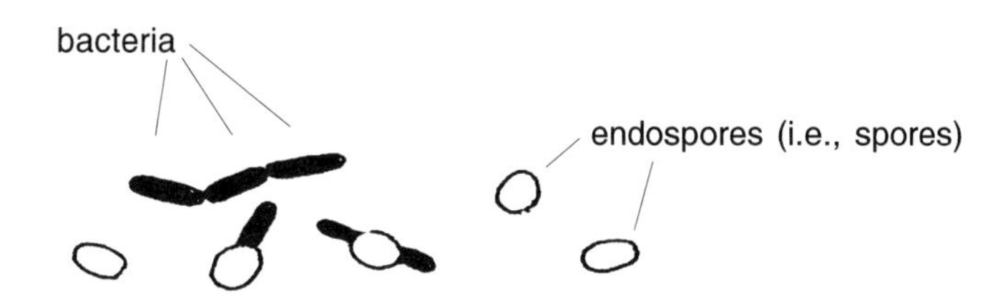

Figure 5.3 *Bacillus* Rod-shaped Bacteria and Oval Endospores

As the most common concern is typically molds, bacterial endospores are often overlooked during indoor air investigations. For this reason, there is has been minimum research and publications regarding this topic. Some have speculated that high levels of *Bacillus* in indoor air is a barometer of past conditions. Either the HVAC system or the building was subject to extreme water damage, saturation, and/or lack of maintenance.

Although some species of *Bacillus* are lethal (e.g., *Bacillus anthracis*), most *Bacillus* bacteria are not pathogenic and are rarely associated with disease. The greatest concern herein is allergenicity.

Thermophilic Actinomycetes

Actinomycetes are filamentous bacteria that resemble fungi in their colonial morphology and production of allergenic spores. Under the microscope, colonial masses appear as thin hyphae (generally much thinner than those found in fungi) with associated spores that are at the low size range for fungal spores.

Ideal temperature preferences range from 37 to 60°C. Yet, these ranges may be, at times, narrow and highly selective. *Thermoactinomyces candidus* grows rapidly between 55 and 60°C but will not grow at 37°C or less. Although most will grow at 37°C, many prefer 45 to 55°C.

Their nutritional requirements are more complex than that of molds, and their spores are more temperature resistant. Thermophilic actinomycetes are the only recognized bacteria that form allergenic spores and can be differentiated from the molds by selective culture media and observation of growth patterns.

The most commonly encountered genera are *Micropolyspora*, *Thermoactinomyces*, and *Saccharomonospora*. Some species of *Streptomyces* have been implicated as allergens as well. In nature, these organisms generally require a nutritionally rich substrate and elevated temperatures. Ideal habitats include moldy hay, compost, manure, and other vegetable matter. Indoor amplification may occur in the heating and humidifying systems where there is also a source of nutrients (e.g., vegetable matter build-up in an air handler with elevated temperatures).

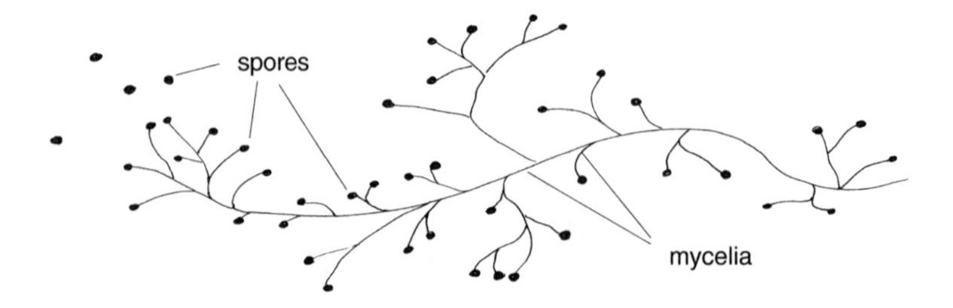

Figure 5.4 *Thermophilic Actinomycetes* are Fungal-like Bacteria

AIR SAMPLING METHODOLOGIES[6]

There are no federal government requirements for monitoring nor are there clearly defined methodologies. Although there have been a few attempts by professional organizations, universities, and private firms to provide guidelines, the most readily accepted guidelines have been set forth by the American Conference of Governmental Industrial Hygienists. Their latest publication, *Bioaerosols: Assessment and Controls*, is frequently referenced herein.

Sampling Strategy

The sampling strategy is subject to the investigator's evaluation of each specific situation. There are a few basic guides to aid the investigator, but they are not hard and fast rules. Careful thought and planning are paramount.

When and Where to Sample

According to the ACGIH, to anticipate high and low exposures, minimum sampling efforts should include a least one, preferably three, sample areas in each of the following areas:

- An anticipated high exposure area (e.g., an area identified as central to health complaints)
- An anticipated low exposure area (e.g., an area identified and confirmed to have minimum health complaints)
- Outdoors near air intakes for the building (e.g., on the roof or along the side of the building where fresh air is taken to supply the indoor area(s) to be sampled)

Other sample sites that should be included are:

- Outdoors near potential sources of bioaerosols that may enter a building (e.g., fresh air entry from open or frequently used doors and windows down wind from a creek bed or waste container)
- Outdoors high above grade and away from potential bioaerosol sources (e.g., background levels not affected by the immediate building environment)

When assessing fungal growth contributions from a ventilation system, locate a site near one of the air diffusers associated with the air handling unit in question. Then take samples at different times during the unit's cycle. Consider the following:

- After the air handling unit has been turned off (generally occurring over a weekend), preferably prior to restart after a weekend of down time
- After the air handling unit has been turned on, restarted after a weekend
- After the air handling unit has been operating for 30 minutes
- During mechanical agitation of the ductwork, preferably when a space is unoccupied and in a fashion to simulate normal maintenance activities or other normal disturbances which might occur to the duct work

Equipment

Although there are several possible choices for sampling equipment, selection is situation dependent. No single sampler can meet all needs. The choice may be a combination of accessibility, reliability, and functionality. Or it may be familiarity.

Slit-to-Agar Impactor

The slit impaction sampler operates by rotating a culture plate below a long, narrow slit, or inlet. The collection substrate may remain stationary, move continuously, or move in increments beneath the slit.

An internal pump draws air through the slit at a flow rate of 50 liters per minute, and the air is impacted onto 10- or 15-centimeter diameter plates with nutrient agar. Sampling duration may be from 1 to 60 minutes.

An advantage to this technique is that slit-to-agar impactors have a wide range of detection and provide limited time-differential information. A disadvantage is that it is not as readily recognized as other samplers and has not been time tested.

Multiple Hole Impactor

The sieve impactor operates by the passage of an air stream through evenly spaced, machine sized holes. The single-stage Andersen impactor has been time-tested and is the most frequently chosen equipment for viable microbe sampling.

A high-volume pump draws air through the impactor at a flow rate of 28.3 liters per minute, or 1 cubic foot per minute. At this pre-set flow rate, particles of a given size (e.g., greater than 0.65 micron in size) are deposited onto the surface of a collection medium (e.g., Petri dish). Faster flow rates will result in the deposition of smaller particles and potentially in loss of viability. Slower flow rates will deposit larger particles. For this reason, it is important that the flow rate be as designed for the most effective collection.

A disadvantage is that the samples are limited in duration. A typical sampling period is from 1 to 5 minutes. Either multiple samples must be taken and averaged over an extended time period (e.g., 8 hours) or random sampling must be accepted as representative of exposures throughout an exposure period. Some professionals have monitored for up to 30 minutes. However, there is a risk of over sampling and loss of viability.

Liquid Impingers

Familiarity and simplicity are the principal advantages to the liquid impingers. They operate by the passage of an air stream through and inertial impaction on a liquid. The liquid may be a sterile solution of water (or surfactant), mineral oil, or glycerol. The latter two retain the viability of the sample more effectively than water. All three liquids minimize dehydration.

Do not be confused into thinking a typical industrial hygiene impinger will work. A high-volume pump draws air through a special all-glass impinger (AGI) at a flow rate of 12.5 liters per minute. The AGI has a pre-established distance between the tip of the inlet jet to the base of the impinger.

The AGI-4 distance to the base is 4 millimeters, and the AGI-30 distance is 30 millimeters. Although a more efficient "particle" collector, the AGI-4 results in greater physical stress because of its shorter distance to the bottom of the impinger. Recovery of viable microbes is more likely with the AGI-30.

Although efficient for collecting a diverse range in particle sizes, this method does not have a high recovery efficiency for hydrophobic bacteria (e.g., *Bacillus*) and some mold spores in water. Recent research, however, indicates effective recoveries in mineral oil and glycerol. Both require considerable static pressure to overcome resistance to air movement through the fluids and filtration is necessary upon receipt by a laboratory. The latter is meeting with considerable resistance due to the difficulty of separating spores down to 1 micron in size from the viscous liquids.

Upon separation, samples can be diluted and plated onto several different nutrient agar. This process allows for different media to be inoculated from the same sample and may permit greater sampling times than the recommended 30 minutes. Yet, more experience is needed with the AGI to reach a level of competency attained by some of the others.

Filtration

Familiarity, simplicity, and long term sampling are the principal advantages to the filtration. Although efficient for collecting a diverse range in spore sizes, sample collection and filter clearance have a severe drying effect on the collected allergens, and analysis results in underestimated spore counts.

Smooth-surfaced filters (e.g., polycarbonate) are less damaging than the coarser filters, and some of the larger industrial hygiene supply companies are developing hydrated filters to overcome the drying effect of long term sampling. The sample airflow rate is 1 to 5 liters per minute.

Sampling duration is variable, up to 24 hours, and air volume may be up to 1,000 liters. The higher flow rates and the longer sampling times will result in even a more extensive loss of spore viability. Filtration is discouraged for the more fragile bacteria.

Centrifugal Agar Samplers

Longer sampling duration and ease of use are the principal advantages to the centrifugal agar sampler. It operates by a rotating drum which draws air at a flow rate of 50 liters per minute with impaction of particles onto the surface of manufacture-supplied agar strips. The sampling duration is up to 20 minutes. Although not all laboratories are familiar and capable of analyzing agar strips, this approach is becoming more widespread.

Sample Duration

Equipment, airflow rates, and culture plate limitations are the limiting factors in sample duration. The most commonly used sampling equipment require a sample duration of 1 to 60 minutes, a small snapshot in the overall exposure time. Airflow rates are pre-set, not subject to change, and culture plates can become overloaded if too much air volume is sampled. The sample duration is the only variable that can be adjusted—knowing the limitations.

Ideally, the investigator wants to collect a minimum number of colony forming units (CFU) with a maximum number of microbial growths per plate. Overloading can render the sample unreadable. Note that some labs will attempt to read overloaded plates and report a greater than number. The ACGIH recommends an optimal collection of 10 to 60 CFU/plate. Yet, in order to stay within this range, the investigator must anticipate the exposures.

For example, with a single-stage Andersen impactor, the lower detection limit would be 35 CFU/m^3 for a 10-minute collection time, and the upper detection limit would be 5,570 CFU/m^3 for a 30-second collection time. Once again, the sample duration is based on anticipated exposures. Professional judgment comes into play!

Sample Numbers

With limited sample durations of less than 10 minutes in most cases and 60 minutes in others, monitoring the entire exposure time requires multiple sampling, which is unfeasible. Thus, a logical, well thought out selection of sampling time(s) is necessary to obtaining results that can be readily interpreted with minimal speculation.

Either the investigator may choose to sample during an anticipated worse case scenario (e.g., respondents on questionnaires state that their symptoms are worse on Monday mornings or after custodial activities), or the investigator may choose to take two to three samples at each site throughout the day. Some investigators may choose to sample every other hour throughout the exposure period. Larger sample numbers result in greater data reliability. Yet, on the practle side, larger sample numbers result in greater expense. The decision on the number of samples to be taken becomes a difficult decision of weighing all factors.

Culture Media

There is considerable controversy as to the appropriate medium to use. Not all molds will grow and create recognizable spore-forming structures on a single nutrient agar. This is one of the single most important and controversial issues in sampling for molds.

General Information

The choice of culture mediaum is dependent upon the organism(s) the investigator seeks to identify and on the laboratory's choice. Keep in mind that there is no single medium upon which all fungi or bacteria will grow. Not all can be cultured, and not all molds will form identifying spores. The failure to form identifying spores is generally reported as mycelia sterilia.

The best, most commonly used culture medium for airborne fungi is malt extract agar. It supports the growth of most viable fungal spores and is an excellent medium for identifying species. Species identification is sometimes important, not only for allergen amplification determination, but for identifying species that may have other effects. For example, *Asergillus flavus* can be deadly for immune-suppressed individuals.

Some media inhibit the growth of undesirable competitors. Rose Bengal agar (RBA) is used for fungi while bacterial growth is kept to a minimum. In some environments where high bacterial levels are anticipated (e.g., agricultural environments), RBA would be the medium of choice.

Stachybotrys molds grow on cellulose agar. Although some laboratories claim this medium is best suited for *Stachbotrys*, some laboratories have demonstrated that Rose Bengal agar works equally well. The trend, however, is toward cellulose agar when *Stachybotrys* mold is the focus.

Bacillus as well as environmental and human commensal bacteria grow well on R2Ac agar with cycloheximide, a fungal suppressant. *Bacillus* and thermophilic actinomycetes will grow on tryptic soy agar (TSA). *Bacillus* and pathogenic bacteria will grow well on blood agar (BA). As different species grow in variable temperature ranges, the choice of medium for *Bacillus* may also be dependent upon the anticipated temperature tolerance for the bacteria under investigation. In an indoor air quality investigation, the most likely *Bacillus* to grow will be that which grows at room temperature. In this case, the R2Ac would be the medium of choice.

The preferred medium for thermophilic actinomycetes is tryptic soy agar. The thermophiles grow best at elevated temperatures as do the pathogenic bacteria. Elevated temperatures tend to kill and/or suppress growth of other organisms.

The thermophilic actinomycetes are incubated at 56°C, and the commensal bacteria are grown at 35°C. All others are grown at room temperature (i.e., 23 ± 3°C).

Special Comments

Plated culture medium can be purchased from a laboratory supply retailer (e.g., Remel or Difco), and some microbial laboratories supply media to their clients. The latter is preferred.

For use in impaction samplers, the plated medium should be evenly distributed in the Petri dish. If it has melted during transportation and is not level upon receipt, do not use the plate. Impaction is based on distance from the air holes to the surface of the agar. If the agar is not flat, the impaction will result in poor sample collection and inconsistent counts.

Culture media dries out after four to six weeks. So, don't overstock. Under stocking may pose a problem as well. Order for anticipated needs with a minimum 10 percent excess. Keep in mind, the plastic dishes may crack in transit, or the investigator may inadvertently contaminate a plate (e.g., sticking a finger into the agar) and require replacement. The perfect world does not exist outside the laboratory.

Equipment must be calibrated to assure adequate flow. This may be up to the discretion of the investigator.. Most investigators do not have adequate equipment for calibration (e.g., large bubble burettes or electronic bubble calibrators are not adequate). The equipment manufacturer generally offers this service and recommends calibration at least annually.

The sampler(s) should be disinfected between sample locations. Isopropanol is the agent of choice at this time. It can be easily purchased and treated, and individual packets are available for purchase.

Excess cleaner (e.g., isopropanol) should be dried prior to its next use, and the sieve holes should be inspected prior to proceeding. Then allow the sampler to run for a couple minutes—at the new location—prior to taking the next sample.

Care should also be taken with the Petri dish cover for the duration of the sampling. A minimum precaution should be to place the cover face down on a clean, smooth surface for the duration of the sampling. To assure a surface is clean, wipe the surface with an alcohol swab prior to putting down the top of the Petri dish.

After the sample has been taken, replace the cover, seal or tape the edges (e.g., laboratory paraffin), label the Petri dish, and store it with the agar side down. Some laboratories prefer ice packs accompany the samples. Check for further instructions.

The laboratory chosen to perform the sample analyses should be consulted prior to each scheduled sample collection for instructions as to their in-house procedures for packaging, shipping, and receiving. If shipment to a laboratory is required, most

Table 5.4 Summary of Culture Media and Anticipated Growth Patterns

Culture Media	Predominant Growth	Incubation Period	Incubation Temperature
Malt extract agar	fungi	1-2 weeks	RT
Rose Bengal agar (suppress bacteria)	fungi	1-2 weeks	RT
Cellulose agar	*Stachybotrys* fungi	1-2 weeks	RT
R2Ac agar	environmental bacteria	48 hours	RT
Tryptic soy agar	Thermophilic actinomycetes and *Bacillus*	48 hours	56°C
Blood agar	pathogenic and commensal bacteria	48 hours	35°C
MacConkey's agar	Gram-negative bacteria and *E. coli*	48 hours	35°C

laboratories require overnight shipments. For samples are shipped on a Friday, weekend, or holiday, special arrangements should be made beforehand. Express delivery services availability and time lags should be determined, and arrangements may be needed with the laboratory for special deliveries.

When choosing a laboratory, consider in-house procedures for sample incubation. Suggested incubation for fungi is 10 to 14 days at room temperature with subsequent identification by genera. Some laboratories will perform a count within 5 to 7 days after receipt. Others feel the slower growing molds will take up to 14 days to grow, and a shorter incubation period may result in incomplete counts and mold identification. The count can be off as much as 10 percent, and an early count may also result in not identifying some of the more important, slow-growing molds (e.g., *Stachybotrys*).

Sample incubation times impact the laboratory turnaround time and completeness of information, considerations which may vary depending on each situation. For example, a quick turnaround is required, and the incubation period for a specific targeted mold type is 5 days. Some laboratories will not assess a sample until the full 14 days have passed. So, in some cases, the investigator may want to locate a laboratory that will evaluate the sample(s) earlier. Some do it routinely and this type of laboratory can be found. See Figure 5.5 for Petri dish growth samples.

Pathogenic molds will not readily grow on most of the media. They grow best at higher incubation temperatures and require much longer incubation periods, up to 3 weeks. If the pathogenic molds could grow at room temperature, on the nutrient media provided, the other fungi would likely overgrow the Petri dish, and newly formed pathogenic mold colonies would not be observed.

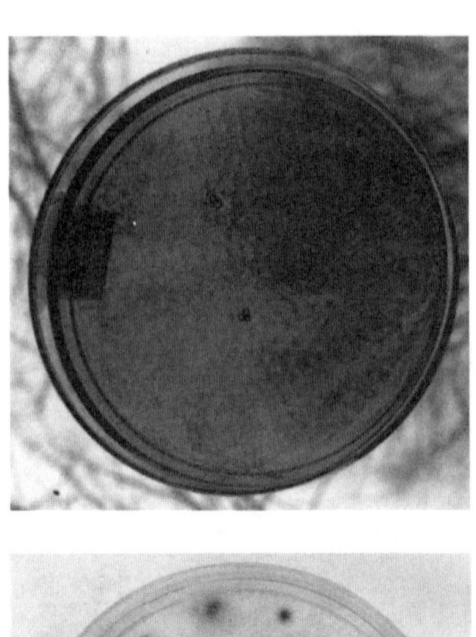

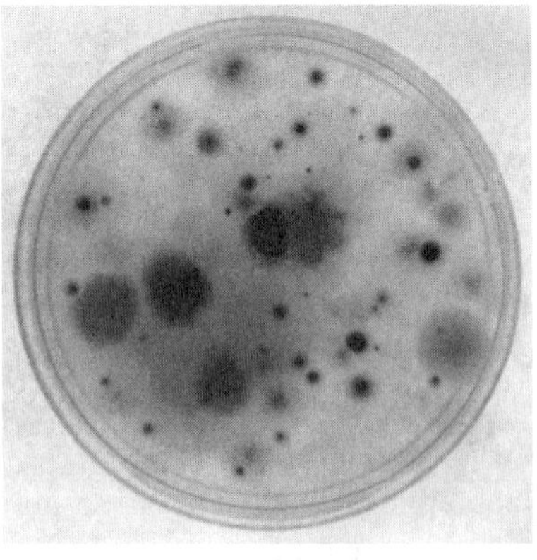

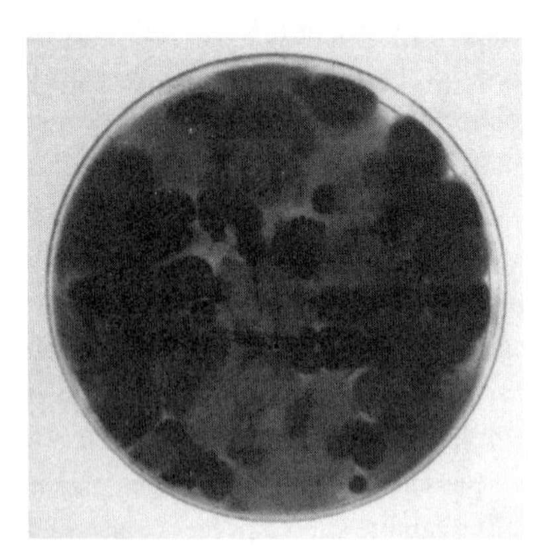

Figure 5.5 Petri dish deposition using impaction samplers. Examples are: (top) impaction marks on agar, (middle) culture with count of 1,761 CFU/minute on malt extract agar after 5 days of incubation, and (bottom) excessive growth on malt extract agar after 5 days of incubation. Contribution by Omega Southwest Consulting, Inc. in Georgetown, Texas.

The recommended incubation for thermophilic actinomycetes is 55°C for 4 days. On the other hand, *Bacillus* bacteria can be incubated within 2 to 4 days along with the thermophiles on the same nutrient medium at 55°C, and *Bacillus* bacteria can be incubated in 2 days at room temperature on the same nutrient media as the thermophiles or on a different medium. The different incubation temperatures are important for identification of all *Bacillus* genera. If *Bacillus* bacteria are amplified and contributing to the total microbial allergens, the most likely *Bacillus* genera to be found amplified indoors would be those that grow at room temperature.

For quality control, some investigators collect a blank along with the air samples. The blank may be an unopened or opened Petri dish. The unopened blank may provide information as to whether the nutrient medium has been contaminated prior to sampling and/or during shipping. The opened blank may give information as to the investigator's sample handling procedures. Not presently a common practice, the submittal of blanks is gaining in popularity.

Procedural Summary

Determine a sampling strategy, when and where to sample. Get the appropriate equipment, and collect the sample(s) with caution. Precautions should be taken not to contaminate media by: (1) properly cleaning the equipment between sample locations; (2) proper management of the sampling media; and (3) proper shipping.

Collect each sample, label containers (e.g., alpha numeric), and log the sample location details as well as air volume sampled or duration of sampling. Additional information the investigator may want to enter is observations (e.g., sample taken immediately under a recently opened ceiling tile), environmental conditions (e.g., relative humidity and temperature), air movement (e.g., directly in line with an air supply diffuser which was blowing at the time of sampling or measured air movement 200 feet per minute at sample location), and odors (e.g., a site identified as having strong mildew odors).

Package all samples in a sturdy container. Some laboratories request special packing procedures (e.g., ship in Styrofoam™ container with ice pack). Ship for overnight delivery to an analytical laboratory.

DIAGNOSTIC SAMPLING METHODOLOGIES

Oftentimes, it becomes necessary to identify the source of amplified microbes. Either remediation of a suspected source has failed to rectify a problem, or the environmental professional chooses to confirm suspect sources at the time of the initial sample taking, possibly after obtaining the initial air sample results and prior to making recommendations. Bulk and surface sampling is the recommended approach.

Sampling Strategy

Purpose is the focus when developing a sampling strategy. Bulk and surface sampling may be performed for any of the following reasons:

- To confirm or deny a suspect source of mold growth
- To identify sources of potential airborne microbes
- To confirm adequate cleanup

If a surface has visible stain, the discoloration may or may not be molds. Surface sampling can be performed not only to confirm or deny its presence but to identify the type of mold. A dark black stain that appears to be a mold on a wall surface may be charred grease or shredded rubber, or a black graphite-like material on the wall may be a specific identifiable mold.

Molds are often found growing on wet fiberglass duct insulation immediately after the cooling coils. This is a common finding, and surface sampling is vital to confirm the presence of molds where expensive remediation may be involved.

Duct-cleaning services and some investigators confirm molds in the relatively dry ducting and use its mere presence to justify cleaning the air ducting. It is no surprise that these same people always do find molds in the air ducts and subsequently require extensive duct cleaning. The investigator should not rely on surface and bulk samples alone, because:

> *... bulk samples cannot replace air samples because the former have not been found to accurately reflect past, future, or even current bioaerosols.* [7]

Once sir sampling has been performed and amplification of certain molds confirmed, the investigator should locate the source. Cleanup can be very expensive, and identification of the source can be vital not only to limit the cleanup but to assure that the actual source has been located and remediated.

After remediation and cleanup of an area, surface sampling may be performed to confirm adequate cleaning. This is a secondary approach after a visual inspection has been performed, and all surfaces appear to be clean.

Where to Sample

Once purpose has been defined, the investigator performs a visual inspection of the area or areas of concern. If microbial growth is apparent, a sufficient number of samples should be taken to represent all suspect components. For instance, if wet duct insulation has a large surface area with thick brown mud-like splotches, a black-green patch, and several white spots, one of each of the three different sites should be taken.

If microbial growth is not apparent, the investigator may either perform several random samples or identify suspect areas based on area activities, interviews, and the presence of moisture/growth media. Custodial personnel or maintenance may have been observed working above ceiling tiles in the complaint area. Carpeting and partition panels may harbor molds but not have a visible presence. Settled water may be contaminated with microbes but not have any visible evidence as to its presence.

When deciding where to sample, the investigator should focus on the following points:

- Microbial growth is influenced by temperature and humidity.
- Dissemination of microbial allergens is influenced by activities (e.g., vacuum cleaners may disperse molds into the air), the presence of molds (e.g., excessive dust collection), air movement over surfaces (e.g., inside air ducts).

Upon confirmation as to the presence of high moisture content in walls, some investigators actually put a hole into the wall in order to take a sample of a suspect microbial growth surface and, upon identification of its presence, spend great sums of money remediating wall spaces that contain mold spores, with no evidence of airborne spores. Although it is possible to have air movement within the wall spaces and entry of wall contaminants into occupied spaces through wall penetrations (e.g., wall sockets), these considerations are rarely tested.

The greater the number of samples taken, the more reliable the statistical end product. Yet, larger sample numbers are more expensive. All factors should be taken into consideration.

What to Sample

Microbes can be found in and/or on almost anything. Thus, the investigator should be prepared to interpret findings prior to sampling. Taking samples without a game plan can result in a conclusion with strange bed fellows.

Samples can be taken of surfaces or suspect bulk materials. Surface sampling may involve loose surface debris and dust, and bulk sampling may involve a substrate and its component debris and dust. The substrate may be a porous solid material (e.g., carpeting or fabric office partition), or it may be a liquid (e.g., water in a drip pan).

Yeasts and molds may grow in flower pots. These are areas often overlooked. Where suspect, soils may also be sampled.

Sampling Supplies

The equipment needed for surface and bulk sampling is less involved than air sampling. Although the equipment can often be purchased at a local grocery or drug store, the analytical laboratory should be consulted prior to obtaining supplies. They may prefer their own sterile supplies or may have developed detailed directions and protocols. Many laboratories will send the appropriate supplies upon request.

Surface sampling is often done by wipe (i.e., swipe) sampling or dust collection. Wipe sampling is performed using a sterile swab that can be returned to a sterile container after a sample is taken. In some cases, the container will have a wetting solution in an ampoule that is broken to release a wetting agent onto the surface of the swab (e.g., 3M Quick Swab). However, some laboratories prefer the investigator use dry swabs.

Dust, debris, and soil can be collected in plastic zip-lock bags, capped vials, or filters (e.g., cassettes) with a suction device (e.g., air sampling pump). Water samples can be collected in capped vials, and moist surface areas may be sampled with a swab.

A sterile template is advisable for swipe sampling. Some laboratories supply a prepackaged plastic 4-inch by 4-inch template. The template provides boundaries within which a sample may be taken and surface area reported.

Bulk sampling requires a sterile container, cutting tool, and latex gloves. Larger samples will require larger containers, and if an abundance of microbes is anticipated, sterile containers may become a mute point.

Procedural Summary

Determine a sampling strategy, where to sample, and what to sample. Get the appropriate supplies, and collect the samples without cross contaminating each new sample. Cross contamination may occur by unprotected hands and uncleaned implements (e.g., template or utility knife).

Collect each sample, label containers (e.g., alpha numeric), and log the sample location details as well as surface area or size/volume of the bulk material. Additional information the investigator may want to enter is visible surface area affected (e.g., visible coverage on approximately 20 square feet), environmental conditions (e.g., wet duct insulation, high humidity, and temperature), and description of stained material (e.g., tea-colored ceiling tiles).

Package all samples so cross contamination, penetrations, and damage cannot occur during shipping. Ship for overnight delivery to an analytical laboratory.

INTERPRETATION OF RESULTS

The interpretation of results is highly controversial. Attempts have been made by various researchers and professional groups to set exposure limits for allergenic spores, but environments, exposure durations, predisposing health conditions, and limitations of viable sampling add fuel to an already heated topic.

Office, agriculture, school, and residential environments will vary considerably in anticipated exposure levels and in tolerance. Anticipated exposures to agriculture workers are quite high, whereas office exposures are generally less than 200 CFU/m^3. See Table 5.5.

Office personnel and agriculture workers are typically healthy adults, and their exposures are limited to workday hours. Yet, even in this group there are people who are more sensitive to allergens than others (e.g., asthmatics). School children have more allergies than adults. They grow out of these allergies as they get older, but they are exposed to all sorts of allergens in their school environments. Residential exposures include infants, elderly people, and adults with debilitating health conditions, and their exposures may be ongoing, 24 hours a day, 7 days a week. Setting one limit to accommodate all is not feasible.

The other problem with establishing an exposure limit for viable microbial allergens is that the count may not be representative of all microbial allergens. All microbial allergens include viable, culturable, and nonculturable microbes. The information derived from viable microbial allergens may be only a small portion of the offending allergens. Although not always applicable, the information derived from viable microbial sampling often needs to be assessed by other means. To date, the most recognized approach to defining environmental problems as they relate to microbial allergens is the assessment of genus variability, a process than has as many approaches as there are investigators.

Table 5.5 Concentrations of Viable Allergens in Agricultural Environments

Workplace Environment	Thermophilic Actinomycetes	Total Fungi
Outside air	10	1,000
Domestic waste management	1,000	100,000[A]
Farming (normal activities)	—	10,000,000
Farming (handling moldy hay)	1,000,000,000	1,000,000,000[B]
Pig farms	1,000	10,000
Mushroom farms (composting)	10,000,000	100
Mushroom farms (picking)	100	100
Sugar beet processing	100	1,000
Cotton mills	100,000	1,000

[A] Mostly *Aspergillus* and *Penicillium.*
[B] Mostly *Aspergillus.*
Excerpted from Crook, B.[8]

Genus Variability

Genus variability involves genus identification and percent of total components. The indoor complaint area is compared to the outdoor area and sometimes to a non-complaint area as well. Where there is a shift in the percent of identified components in the indoor air verses the outdoor air, indoor amplification (or growth on a building material) is indicated. A shift in the percent is interpreted several different ways.

The most common approach is to assess percent of total microbial allergens. For instance, the outside air contains 85 percent of *Cladosporium*, 10 percent of Alternaria, and 5 percent *Penicillium*, and the indoor air contains 90 percent *Penicillium*, 8 percent *Cladosporium*, and 2 percent *Alternaria*. This example, involving unrealistically limited numbers of mold spores in a total count, is a good indicator that *Penicillium* molds are growing indoors.

Some investigators generate a graph of the more realistic 10 to 12 identified molds and compare these. This approach is not as easy to assess as where the top 2 or 3 identified molds are compared without the complex jumble of all molds.

Some investigators cut do not perform a genus variability assessment where the total count is low (e.g., less than 100 CFU/m^3). A low colony count is subject to considerable variablity. The collection of only a few spores in a small air volume can vary a lot due to sampling and analytical errors. The statistical variability due to sampling and analytical errors has yet to be published for these viable mold sampling methodologies.

Amplification is often considered where certain toxigenic molds are identified indoor, irrespective of that identified outdoors. *Stachybotrys* is one such mold. Be wary, however, of declaring amplification where there is a single colony of a given mold. A single colony may be an anomaly that occasionally occurs in outdoor air samples as well.

The assessment may also be limited by the laboratory reporting techniques. Most laboratories identify the more prevalent genera, up to a limited number (e.g., five of the most numerous mold colonies). Some identify all recognizable genera on the basis of growth structure and patterns, whereas others identify the most prevalent genera. Many will attempt to identify species of *Aspergillus* as well (e.g., *Aspergillus flavus*). Other genera must be re-plated for species determination. This involves more expense and culturing time (e.g., an additional 2 weeks). A colony growth may be declared as unidentifiable, or it is referred to as mycelia sterilia. The latter means the mold is sterile, does not for fruiting bodies/spores, in the nutrient medium provided. These (mycelia) can, however, be replated onto other media where they may grow and potentially be identified. Due to recent concerns, most laboratories will also identify one of the species of *Stachbotrys* (e.g., *Stachybotrys chartarum*).

The same interpretation process should apply for thermophilic actinomycetes and *Bacillus* bacteria as well. Yet, separate sample culture media and analytical techniques are required for the bacteria.

Prior to 1999, ACGIH publications recommended numeric exposure limits. These have since become obsolete, but some investigators persist in demands for airborne exposure limits.

Airborne Exposure Levels

Reports indicate that outdoor spore counts routinely exceed 1,000 counts/m^3 and may average near 10,000 counts/m^3 during warmer, more humid months. In some parts of the country, the outdoor levels may exceed 20,000 counts/m^3. The levels vary throughout the day and only a percent of these are viable.

With rare exception, indoor levels of allergenic molds that are less than 100 CFU/m^3 are not typically associated with complaints. A couple exceptions are special environments, such as hospitals, where there are immune suppressed patients that must avoid any level of certain molds (e.g., *Aspergillus flavus*) and people with life threatening respiratory problems.

Some environmental professionals chose 200 CFU/m^3 viable molds as the minimum indicator for probable indoor exposure limits in environments with a relatively healthy adult population (e.g., office buildings). The previous ACGIH recommended limits were even more permissive.

In 1986, the ACGIH recommended a limit of 10,000 CFU/m^3 total fungi, 500 CFU/m^3 for one genus of mold, and 500 CFU/m^3 for thermophilic actinomycetes. In 1989, they recommended comparative sampling and stated that the presence of thermophilic actinomycetes, typically associated with agricultural environments, is sufficient to indicate contamination.

While comparison sampling has become a sound scientific approach for ascertaining amplification, viable sampling provides only a piece of the picture. An easily referenced standard for acceptable mold and bacteria spore levels is not just over the horizon.

Bulk and Surface Sample Results

The laboratory reports bulk samples in terms of identified viable microbial numbers per weight or volume of material. This is usually in terms of CFU/gram or CFU/milliliter.

Surface sample results are reported based on the identified viable microbial numbers per surface area. This is typically done in terms of CFU/cm^2 or CFU/$inch^2$.

Although the numbers may appear quite high to an investigator (e.g., 2,000,000 CFU/gram), there are presently no guidelines to assist in the assessment of these levels. Poor correlation exists between bulk and surface sample results. The best use for the results is identification of probable sources that result in amplification of allergenic microbes indoors.

For example, if there was 43 percent *Aspergillus syndowii* indoors and none in the outdoor air sample and 100 percent *Aspergillus syndowii* in a bulk sample, the bulk sample location is immediately suspect of being the source or one of the sources of amplified airborne mold spores. If the bulk sample had contained *Aspergillus versicolor* (and was not found in the air sample), the sample site would not be suspect. This is where species classification can be helpful.

Although they may have the same mold types, surface samples will typically be in smaller numbers. For instance, duct insulation may have 2,140,000 CFU/gram (and 100 percent *Penicillium*) with a surface level of 2 CFU/cm^2. An attempt to interpret the numbers can be a daunting task. However, it is clear that most of the contaminant is deep within the porous surface of the duct insulation.

HELPFUL HINTS

When taking short-term samples, either use a stop watch, a timer with a second hand, or an in-line timer that will turn off the sampler. One- and 2-minute samples that go over by 10 or 15 seconds can make a big difference in the calculated air volume and final results if not reported. If you go over, either report the air volume on the basis of the time sampled in terms of minutes and seconds or start over.

It is also distinctly possible, inevitable, that the investigator will at some time poke his/her finger in the sterile nutrient agar. Some laboratories ask that these sites be marked. On the other hand, don't use that plate. Start over again.

As a level of confidence is reached, the investigator will typically choose one sampler and stay with it. While this is a prudent practice, other methods that may meet more expanded needs are constantly being improved and/or refined. Many approaches are tried and true, but new technology emerges every day. The investigator may follow the progress of new technology through professional journals, conferences, and publications.

SUMMARY

Viable microbial allergens include molds and spore-forming bacteria. Yet, viable is the key word. All airborne microbial allergens may not grow on the nutrient media used for air sampling. For this reason, a count is often questioned. The primary information the investigator needs to target is the comparative sampling (i.e., comparison of complaint, non-complaint, and outdoor samples). Then, too, if the sampling strategy and methodologies are poorly planned and executed, the results can be difficult, if not impossible, to assess.

Diagnostic sampling can be invaluable if performed methodically with a clear concept of what one is looking for. Sampling for sampling's sake will go nowhere.

Viable microbial sampling is best used in conjunction with pollen/fungal spore counts. In one situation, the fungal spore counts had an elevated anomaly they wanted to call *Aspergillus/Penicillium*. Two weeks later as the viable results became available, the count was low and the anomaly turned out to be oil droplets. Viable sampling does have a place in obtaining information.

REFERENCES

1. ICMSF. *Microbiological Ecology of Foods*. Academic Press, New York (1980).
2. ACGIH. *Biioaerosols*. ACGIH, Cincinnaati, OH (1999). p. 19-6.
3. Crissey, John T., MD, H. Lang, and L. Parish. *Manual of Medical Mycology*. Blackwell Science, 1995, pp. 190-201.
4. Al-Doory, Yousef and J. Domson. *Mould Allergy*. Lea & Febiger, Philadelphia, Pennsylvania, 1984. pp. 36-37.
5. Jaeger, Deborah L., M.S. et.al. Microbes in the Indoor Environment. PathCon Laboratories, Norcross, VA (1998).
6. ACGIH. *Biioaerosols*. ACGIH, Cincinnaati, OH (1999).
7. ACGIH. *Biioaerosols*. ACGIH, Cincinnaati, OH (1999). p. 12-3.
8. Crook, B. and J. Lacey. Airborne allergic microorganisms associated with mushroom cultivation. Grana. 30:445 (1991).

Chapter 6

PATHOGENIC MICROBES

Pathogenic microbes of the greatest concern in indoor air quality investigations are disease-causing fungi and bacteria to which building occupants are most likely exposed. Although a proactive approach to problem identification is desirable, exposures to pathogens and toxic microbes are a rare concern unless a large number of associated illnesses are reported.

Sampling requires lengthy culture times, complicated by the presence of other environmental, nonpathogenic microbes that may mask and prevent identification of the pathogen by overgrowing the culture medium. Then, when a suspect colony is isolated, additional time is required to perform more extensive, complicated analyses/cultures in order to confirm its identity and, in many cases, to determine the species and strain.

Whereas many of the pathogens are potentially lethal, the time required for identification is frequently not practical. For this reason, the presence of a pathogen is often assumed where patients are symptomatic, reacted upon later. Typically, health professionals attempt to isolate the source environment through studying sick patients' habits and environmental associations, trying to recreate a common denominator for an observed epidemic.

Although pathogenic diseases are frequently occupation and area dependent, the greatest attention to pathogen spread occurs in hospitals. An infected patient aerosolizes (e.g., coughs or sneezes) viable pathogens, and immune-suppressed patients are particularly susceptible aerosolized pathogens and nonpathogens. Immune-suppressed patients include those with AIDS, organ transplants, chemotherapy treatments, and diabetes. Operating rooms and invasive diagnostic testing areas are possibly involved. Infants and the elderly can be compromised.

Those weakened by living conditions (e.g., the homeless) and inadequate nourishment (e.g., Somalians) have an enhanced susceptibility to pathogens. Then, too, some pathogenic microorganisms are capable of causing disease in healthy individuals under special circumstances. Contaminated intravenous solutions and hospital implements provide an easy avenue for entry. Puncture wounds in contaminated environments are fertile soil, and damaged tissues propagate invasive challenges. Opportunity for pathogenic assaults is subsequently dependent upon the environment.

Some pathogens multiply in air handlers and water reservoirs. Many are contained within environmental dust. The problems are rarely identified prior to the spread of disease to others associated with the dispersing mechanism. Sampling is generally requested after a problem has arisen. Proactive sampling in environments where specific pathogens may thrive is indicated.

As invasion and amplification require viable pathogens, the best sampling is for viables. However, viable samples require lengthy culture times, long periods of time not always prudent in a potentially lethal environment.

Methods are available which can identify some pathogens rapidly and accurately. However, these latter methods identify nonviable and noninvasive microbes. They are limited, and the results are sometimes difficult to interpret. Yet, the information is provided herein should an investigator require rapid processing until more reliable information can be provided at a later date.

AIRBORNE PATHOGENIC FUNGI[1-3]

Pathogenic fungi are those fungi that cause disease through inhalation of airborne spores (e.g., *Histoplasma capsulatum*) as well as potentially causing disease in immune-suppressed patients and individuals with skin surface damage (e.g., *Aspergillus flavus*). Skin surface damage may include wounds, open sores, cuts, abrasions, burns, and dry cracks. Symptoms of fungal disease range from localized, surface discomfort to whole body invasion, occasionally death.

The primary focus of this section is to discuss the airborne fungi that can invade and cause death. Only the principal pathogenic molds are discussed. Those that are infrequently or not typically found in the United States may require additional research, strategy, and sampling development. The following information should provide direction and guidance to addressing potentially lethal environmental situations.

Disease and Occurrence

Understanding the diseases caused by specific pathogenic fungi and regions/areas where they primarily occur is necessary to determine a need for sampling. At the same time, however, the environmental professional must not be limited by this information. An outbreak may be occur when least expected.

Aspergillus

Several species of *Aspergillus*, primarily the *Aspergillus fumigatus* and *flavus,* can cause a disease commonly referred to as aspergilliosis. As the genus *Aspergillus* is one of the more common fungal spores found in air monitoring, the pathogenicity is minimal, unless an individual is immune-suppressed or has a debilitating illness. The initial symptoms may be localized to the lungs, ears, or perinasal sinuses. Once the fungus has found a place to grow, the hyphae may grow into the bloodstream to deposit spores to be disseminated to other parts of the body. The result may be the formation of abscesses or granulomas in the brain, heart, kidney, and spleen.

As *Aspergillus fumigatus* and *flavus* thrive in immune-suppressed patients, the greatest areas of concern are in the hospitals, clinics, nursing homes, and hospices. Immune-suppressed patients include AIDS patients, organ transplant patients (e.g., heart transplants), premature infants, and radiologically-treated leukemia patients. Invasive diagnostic suites and operating rooms are particularly critical. The presence of even low levels of the invasive forms of certain *Aspergillus* species can be deadly. The hospitals seek to control these environments. See Table 6.1 for historic outbreaks.

Some studies have included several fungal species to the list of potential invaders. One such study includes the following:[4]

- *Aspergillus fumigatus*
- *Aspergillus flavus* (also produces the carcinogen aflatoxin)
- *Aspergillus niger*
- *Aspergillus terreus*
- *Petriellidium boydit*
- *Fusarium spp.*
- *Mucoraceae spp.*
- *Phoma spp.*
- *Alternaria spp.*
- *Penicillium spp.*

Other studies include *Zygomycetes* and *Rhizopus*.[5,6] Most agree, despite the failed concurrence of opinions, that *Aspergillus fumigatus* and *flavus* are the predominant fungal spores of concern in hospitals. The other species should be considered potentially invasive by the environmental professional.

There are a few occupational exposure concerns regarding the genus *Aspergillus* as well. Aspergilliosis was first described by an Italian in the 1700s as an occupational disease of people who handled grains and birds.

This concern, however, has lost its significance in modern times. As the fungus grows best on compost and damp organic matter, farmers and field workers are at a potential risk for high airborne exposures to *Aspergillus*.

Histoplasma capsulatum

Histoplasmosis, caused by inhalation of the mold spores of *Histoplasmoa capsulatum*, results in a variety of clinical manifestations. Primary histoplasmosis is a common, region oriented, benign disease of the lungs involving a mild or asymptomatic pulmonary infection with a cough, fever, and malaise. Chronic cases may go on to experience a productive cough, low-grade fever, and a chest X-ray showing

Table 6.1 Historic Outbreaks of Nosocomial Aspergillosis

Number of Patients	Type of Disease	Host	Environmental Factors
7	--	bone marrow transplants	Recent construction and defective air conditioner
32	pneumonia/sinusitis	respiratory patients	Building construction and devective AHUs and colonization
10	invasive pulmonary disease	immunosuppressed (7) malignancy (2) elderly (1)	road construction and defective air conditioners
9	invasive pulmonary disease	--	building construction
3	invasive pulmonary disease (2) and colonization (1)	renal transplants	hospital renovation
8	invasive pulmonary disease with one dissemination	acute hematological	wet fireproofing and dust on false ceiling; malignancy, neutropenia
4	invasive pulmonary disease with one dissemination	renal transplantation immunosuppressive Rx	pigeon excreta external through the air intake

() Numbers in parentheses are host breakdown of host patients from first column.
Excerpted from *Nosocomial Aspergillosis: An Increasing Problem.*[7]

cavitation. As the cavitation resembles tuberculosis, however, many cases are misdiagnosed. Then, a small percent (less than 1 percent of all cases) of those showing symptoms develop in the progressive form, which can be lethal.

Progressive, systemic histoplasmosis is less common. It is characterized by emaciation, leukopenia, secondary anemia, and irregular pyrexia. There is frequently ulceration of the naso-oral-pharyngeal cavities and intestines with generalized infection of the lymph nodes, spleen, and liver.

There have been reports of *Histoplasma capsulatum*-related chronic meningitis in non-immune-suppressed hosts, and some have speculated that the fungus also contributes to persistent or recurrent carpal tunnel syndrome.[8,9] Symptoms in immune-suppressed patients are highly variable, and the fungus is generally associated with other pathogens.

Within the United States, *Histoplasma capsulatum* is endemic in the Central Mississippi Valley, Ohio Valley, and along the Appalachian Mountains. See Figure 6.1. Other endemic areas include sections of South America and Central America, and there are occasional outbreaks in various parts of the country and world. It is thought that the spores become airborne, and individuals become infected by inhaling *Histoplasma capsulatum* from the air. Disease severity in a healthy adult is generally dose-related.

The growth of the fungus appears to be associated with decaying or composted manure of chickens, birds (especially starlings), and bats (e.g., "cave disease"). Thus, the potential for elevated exposures occurs predominately around the following areas:

Figure 6.1 Map of United States depicting the states/areas where the occurrence of "Histoplasmosis" is known.

- Soil in and around chicken houses or silos that have not been disturbed for a long time
- Soil under trees that have served as roosting places for starlings, grackles, or other birds
- Farm land where fertilizers or soil-containing chicken droppings or large quantities of organic matter have been spread
- Demolition dust in and around a building where pigeons had previously roosted

Contaminated soils and dust are the greatest sources of exposure to the drought resistant spores. Dry, dusty, windy environments are the most likely to spread the airborne spores.

Coccidioides immitis

Coccidioidomycosis (often referred to as "Valley fever"), a disease caused by inhalation of the mold spore of *Coccidioides immitis*, is generally an acute, benign, and self-limiting respiratory infection. It is characterized by chills, fever, cough, and localized chest pain. Less frequently, the disease involves the visceral organs (e.g., bone, skin, and subcutaneous tissue) where abscess may develop. Illness usually lasts for two to three weeks, leaving some scarring of the lungs. However, if an individual has had a previous sensitizing exposure, the symptoms can be lethal.

In the more advanced stages of the disease, abscess formation in the lungs and the rest of the body, including the central nervous system, may occur. Immunologic resistance generally develops in those who recover.

Organ involvement in immune-suppressed patients includes the skin, bones, joints, and central nervous system. It may or may not be associated with other pathogens.

The disease is endemic in hot, arid/semiarid areas of the southwestern parts of the United States, northern Mexico, and South America. Endemic regions in the United States include sections of Arizona, California (e.g., San Joaquin Valley), Nevada, New Mexico, Texas, and Utah.[10] See Figure 6.2. The fungal spores become airborne, and disease is thought to be dose-related.

Contaminated dust is the principal source of exposure. Therefore, workers and households in arid or semiarid regions have a higher risk of contracting the disease than others. Most of the infections occur during the dry season, following dust storms. Occupations of greatest risk include migrant workers, farmers, construction workers, military personnel, and road workers. Clinical laboratory exposures may also occur due to inadequate controls of the highly infectious, amplified cultures.

Sampling should be in dusty occupational environments, in regions endemic with *Coccidioides immitis*. With a few minor exceptions, the sampling methodology for this fungus is similar to that of *Histoplasma capsulatum*.

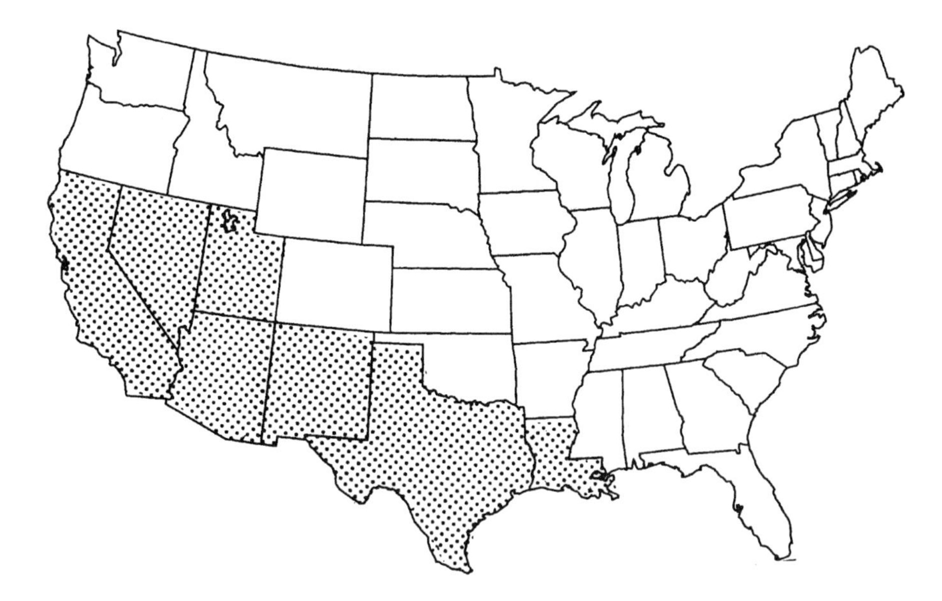

Figure 6.2 Map of United States depicting the states/areas where the occurrence of "Coccidioidomycosis" is known.

Cryptococcus neoformans

Cryptococcus neoformans is a yeast that attacks the central nervous system, producing a subacute or chronic meningitis. It may also impact the lung tissue and cause low grade inflammatory lesions which may be mistaken for tuberculosis or a neoplasm. Occasional cases arise whereby the fungus gives rise to localized nodular lesions of the skin or generalized infections with lesions of the skin, bones, and viscera.

Rare in healthy adults, *Cryptococcus neoformans* varieties *neoformans* and *gattii* are reported to be the cause of death in 6 to 10 percent of the HIV patients. Thus, they are the most common AIDS-related, lethal mycoses.[11] These strains have been known to cause blindness in AIDS patients and are the foremost cause of central nervous system infections in the same. Diabetics, cancer patients, and recipients of organ transplant patients are at risk of generalized infection. Fungal infections range from 5 percent in recipients of kidney transplant patients to as high as 40 percent among recipients of liver transplants.[12]

The most commonly reported source of contaminated dust is pigeon droppings. There are also numerous reports from Australia that the source of infections can be attributed to certain species of Eucalyptus (e.g., River red gum and Forest red gum).[13]

Prior to an incident or outbreak, routine sampling for fungi, identifying *Cryptococcus neoformans,* is indicated especially in areas where immune-suppressed patients may be compromised the greatest (e.g., operating rooms). Particular attention should be given to the air handler systems (e.g., growths occurring beyond the filtration devices) supplying the operating rooms, minor surgery suites, and invasive diagnostic rooms.

Cryptococcus neoformans occurs in cultures in the form of round yeast-like cells, surrounded by a large gelatinous capsule which is thought to contribute to its ability to resist phagocytosis and bypass the initial body defenses. Although they measure 5 to 20 microns in diameter when in tissues, the cells measure 2 to 5 microns in culture media.

Other Pathogenic Fungi

Most of the other fungi which have not been mentioned are either nonpathogenic, rare occurrences of disease in the United States, or rarely infectious through inhalation. For easy reference, potentially opportunistic and rarely pathogenic airborne fungi are identified in Table 6.2 along with other health effects of the more common airborne fungi. This information is intended to provide a quick easy reference, not to burden the reader with scare tactics. Should the occasion arise that a common mold is suspect of causing disease, additional research is recommended.

Sampling and Analytical Methodologies[14,15]

Although many are drought resistant, most pathogenic fungi do not retain their viability during air sampling or may be overgrown on the nutrient media by other fungi, masking the presence of the slower growing pathogens. However, where air sampling has been performed for allergenic viable fungi, a pathogen may inadvertently become identified. The methodologies for air sampling for allergenic molds are detailed in Chapter 5, "Viable Allergenic Microbes." The most commonly identified pathogenic fungus found during routine viable allergenic mold sampling is *Aspergillus*.

Aspergillus fumigatus is thermotolerant and will grow at temperatures of 45°C, or higher, which is an extreme temperature for most allergenic fungi. Culturing a viable allergenic mold sample at this temperature serves as a means for limiting other fungi from growing and overgrowing the *Aspergillus fumigatus*. On the other hand, the least likely to be identified during viable allergenic mold sampling are *Histoplasma capsulatum*, *Coccidioides immitis,* and *Cryptococcus neoformans*.

The method for *Histoplasma capsulatum* and *Coccidioides immitis* is specific for these pathogens only. Sampling involves the collection of settled dust from surfaces and placement in a sterile container. Yet, due to the analytical costs, proactive sampling is rarely performed. In one recent laboratory quote, the minimum analytical cost was $4,000. The reason for the high cost is that analysis involves injecting

Table 6.2 Health Effects of the More Common Airborne Fungi [16-19]

Fungus	Allergenic	Pneumonitis	Hypersensitivity Pathogenic	Potentially Mycotoxin(s)	Possible Comment(s)
Acremonium	X	Humidifier Fever	rare	antibiotic	require high moisture
Alternaria					grows on cellulose
Arthrinium	X	—	potentially oppor.	—	—
Aspergillus					some xerophilic species
A. fumigatus	X	X[A]	Aspergilliosis & oppor.[B]	X	grows at 37°Cor higher
A. flavus	X	X[A]	Aspergilliosis & oppor.[B]	X	grows at 37°Cor higher
A. niger	X	X[A]	swimmer's ear	X	—
A. versicolor	X	X[A]	—	X	—
A. sydowii	X	X[A]	potentially oppor.[B]	X	—
Aureobasidium	X	Humidifier Fever	— sauna taker's lung	—	moist bathrooms, under window sills, liquid waste
Bipolaris	X	—	X	X	—
Botrytis	X	X	—	—	"noble rot" wine grapes, damages plants
Chaetomium	X	—	rare	X	damp sheetrock paper
Cladosporium	X	Hot Tub Lung	*Cladosporium carrionii* Moldy Wall Hypersensitivity	X	growth range: 0 ° -32°C moisture: >0.85
Curvularia	X	—[C]	rare	—	—
Epicoccum		X	—	—	antibiotic growth range: 4°-45°C moisture: >0.86

Table 6.2 Health Effects of the More Common Airborne Fungi (continued)

Fungus	Allergenic	Pneumonitis	Hypersensitivity Pathogenic	Potentially Mycotoxin(s)	Possible Comment(s)
Fusarium	X	—	X & oppor.[B]	X	moisture: > 0.86[D]
Geotricum	X	—	rare	—	—
Mucor	X	X	rare	—	moisture: >0.90[D]
Nigrospora	X	—	rare	—	—
Penicillium					moisture: > 0.78
P. marneffei	X	X	rare	X	—
P. notatum	X	X	—	antibiotic	penicillum
P. griseofulvum	X	X	—	antibiotic	griseofulvum
P. brevicompactum	X	X	—	X	—
P. fellutanum	X	X	—	X	—
P. viridicatum	X	X	—	X	—
Rhizopus	X	X	rare	—	moisture: > 0.94[D]
Stachybotrys	X	—	—	X	moisture: > 0.94[D] grows on cellulose
Torula	X	—	—	—	grows on cellulose
Trichoderma	X	X	potentially oppor.[B]	X	grows on cellulose
Ulocladium	X	—	—	—	—

-- Either not reported, not confirmed, or not known.

A Humidifier fever, malt worker's lung, wood trimmer's disease, straw hypersensitivity, and farmer's lung.

B Opportunistic with weakened conditions and immune-suppressed individuals.

C Allergic fungal sinusitis.

D Grows at 37°C or higher.

the sample(s) into laboratory rodents, organ extraction, and culturing the isolated organism, and the laboratories require at least ten samples be handled in like fashion. This is included in the minimum charge. Another limitation to this process is time. The laboratory turnaround is at least two months. Given all the methodology limitations to identifying *Histoplasma capsulatum* and *Coccidioides immitis* in occupied spaces, sampling is generally performed only where disease is already known. Where the extent and impact of the associated disease is increasing, and corrective actions have been initiated, the investigator may take a set of samples prior to corrective controls and another set after remediation in order to confirm adequate measures have been taken. Once again, the time delay is unavoidable, and by the time the results are known, the building will most likely have long since been reoccupied.

Cryptococcus neoformans also requires a dust sample and is difficult to identify. As with *Histoplasma capsulatum*, sampling involves the collection of settled dust from surfaces and placement in a sterile container. After sample collection, the method varies. Only one sample is required. It is serial diluted, plated on special nutrient agar at the laboratory, and cultured at 37°C. Once identified, species and sometimes variety (e.g., *Cryptococcus neoformans,* variety *neoformans* and variety *gattii*) determinations require additional plating (e.g., slant specialized nutrient agar in test tubes). This requires a trained laboratory technician and may involve several steps with additional culturing, plating, replating, and stains. These additional efforts may also be time consuming and expensive.

Interpretation of Results

There are no definitive guidelines for interpretation of results. Where a pathogenic fungus is region dependent, baseline levels in symptom-free areas in endemic regions may provide a means for dose comparisons. Presently, there are no widely published findings nor recognized research studies which have identified baseline levels and suggested response limits.

For immune-suppressed patients, any level of exposure may be potentially lethal. Thus, its confirmed presence in a hospital treatment area or association with medical implements should constitute reason for concern. The source should be found and remedied, and medical implements cleaned. With immune-suppressed patients, there has been no determination as to an acceptable level. Many health/environmental professionals, when assessing sensitive environments, express concern at any level of identification and remediate to zero detection

AIRBORNE PATHOGENIC BACTERIA[20-22]

Most airborne bacteria are nonpathogenic. The most numerous bacteria are the environmental bacteria, whereas indoor bacteria that are associated with humans are referred to as commensal bacteria. Pathogenic bacteria are not common and are sometimes difficult to isolate.

Most of the pathogenic bacteria indoors occur when they are aerosolized from soil, water, plants, animals, and people, but outdoor exposures may also occur. The latter are generally associated with windblown dust. Airborne pathogenic bacteria usually infect the respiratory tract.

Enormous numbers of moisture droplets are expelled during coughing, and smaller amounts are expelled through talking. A single sneeze by an infected person may generate as many as 10,000 to 100,000 bacteria. On the more positive side, most bacteria do not survive once they have become airborne.

Each of the pathogenic bacteria has its own distinct survival adaptation that permits it to survive long enough to invade its target host. The bacterium *Legionella* is highly resistant to acids. The thicker walled, Gram-positive bacteria (e.g., diphtheria) are resistant to drying, and pathogenic *Bacillus anthrax* forms protective spores. The more prominent airborne pathogenic bacteria found within the United States, those posing the greatest concerns in the occupational and indoor air environments, are discussed herein.

Pathogenic *Legionella*[23,24]

Diseases caused by the bacterium genus *Legionella* are estimated by the Centers for Disease Control at between 25,000 and 100,000 per year in the United States alone. Collectively, these diseases are referred to as "legionellosis" (e.g., Legionnaires' disease and Pontiac fever).

There are thirty-four known species of *Legionella*, fifty known serogroups. Although many of the species have not been implicated in human disease, *Legionella pneumophila*, Serogroup 1, is the most deadly, most frequently implicated form associated with Legionnaires' disease. The latter Serogroup 1 and many of the others are frequently found in all environments and water reservoirs. Their presence alone is not sufficient to constitute a threat.

Virulence is related to the following:

- Species and serogroup
- Total amount of viable bacteria in a water reservoir
- Aerosolization of contaminated water
- Distribution of aerosolized droplets to human hosts
- Cooling towers and evaporative condensers—least potential
- Water heaters and holding tanks

- Pipes containing stagnant water
- Faucet aerators
- Showerheads
- Whirlpool baths
- Humidifiers and foggers—highest potential
- Reduced host defenses

Other sources of *Legionella* have been identified in unique environments. In one case, an outbreak was tracked back to the misting of produce in a Louisiana grocery store.[25] Another involved a five-patient outbreak that was linked to the aerosol from a decorative fountain associated with a private water supply in an Orlando, Florida hotel.[26]

Warm water reservoirs provide the greatest potential for amplification of the viable organisms. Viable organisms have also been isolated from stagnant pools, lakes, and puddles of water.[27] An example of a highly virulent situation is the presence of *Legionella pneumophila* in large amount, aerosolized and distributed from a warm humidifier into the living environment of an immune-suppressed, elderly patient.

Although Legionnaires' disease is primarily a disease of the elderly, it may impact younger patients whose health has been compromised and immune-suppressed patients (e.g., AIDS patients). In a healthy adult of working age, infections are usually asymptomatic and may in some cases result in mild symptoms of headache and fever. In the elderly, prior to the onset of pneumonia, intestinal disorders are common, followed by high fever, chills, and muscle aches. These symptoms progress to a dry cough and chest/abdominal pains. Where death occurs, it is typically due to respiratory failure. Death occurs to only fifteen percent of the less than five percent of the exposed individuals showing disease that develops within three to nine days of exposure.

Pontiac fever is not quite so pathogenic. Within two to three days of exposure, approximately 95 percent of all exposed individuals develop short-term flu-like symptoms.

The *Legionella* bacterium is a thin, Gram-negative rod with complex nutritional requirements. A unique characteristic that is used in isolating the genus is its ability to survive at pH 2.

Sampling and Analytical Methodologies

Legionella pneumophila is fragile. Once aerosolized and dehydrated, it loses its viability. For this reason, air sampling is unreliable, not recommended. Microbe damage, and loss of viability, may result in diminished counts.

Water sample collection procedures have been developed. They are considered more reliable and a means for interpreting the results has been developed. Whenever possible, one liter of water should be collected in a sterile container. Screw-capped, plastic bottles are preferred collection containers. If the water source has recently been treated with chlorine, neutralize the sample(s) with 0.5 milliliters of 0.1 N sodium thiosulfate. Several samples should be taken from suspect/potential reservoirs.[28]

Sometimes, a liter of water cannot be collected. Collect as much as is feasible, and place it in an appropriately sized container. A one-milliliter sample may evaporate from or become irretrievable from a one-liter container. Where sampling from a faucet or showerhead reservoir, available water is minimal. In such instances, the reservoir should be swabbed with sterile swabs (e.g., polyester medical swabs with a wood stick), contained within a sterile enclosure. Upon sampling, each sample swab should be submerged in a small amount of water from the source outlet (e.g., shower, faucet, etc.).

Upon completion of sample collection, all samples should then be sent by overnight freight to a laboratory for analysis. If this is not possible, the samples should be refrigerated until they can be processed. Otherwise, avoid temperature extremes both in storage and shipping.

At an experienced laboratory, the known aliquots of collected water sample will be plated onto nutrient media and cultured for up to three weeks. The nutrient medium of choice is enhanced, buffered-charcoal yeast extract agar. Prior to the plating, many of the samples are acid-treated for 15 to 30 minutes and then neutralized. Most of the other non-*Legionella* microbes are destroyed, and the *Legionella*, having retained its viability, grows unrestricted.

The plates are incubated at 35°C and examined daily for ten days. *Legionella* bacteria are slow growing organisms, and colonies that develop early are not likely to be *Legionella* bacteria. Sometimes, where there are excessive growths, a determination may be made within a few days. Otherwise, the growth may not appear for three weeks. Results are reported in terms of organisms per milliliter of water.

Interpretation of Results

Exposure action levels for contaminated water have been developed by researchers. Proposed limits are based upon extensive experience and field studies. The researchers determined the levels of organisms present in various water sources where Legionnaires' disease was known and where it was not a reported problem. The data was compiled and numbers compared. It is from these studies that the action levels were created in order to anticipate and avoid costly outbreaks. These suggested limits are provided within Table 6.3. They are intended for use as "guidelines only."

The action levels are 1 through 5. Action Level 5 requires the greatest amount of care (e.g., immediate cleaning and/or biocide treatment). Each suggested action includes the preceding actions as well. The minimum recommendation for Action

Table 6.3 Suggested Remediation Action Criteria Levels for *Legionella*

Legionella Organisms per Milliliter	ACTION LEVELS CT/EC	Potable Water	Humidifier/Fogger
< 1	1	2	3
to 9	2	3	4
to 99	3	4	5
to 999	4	5	5
> 1,000	5	5	5

CT/EC: Cooling toweers and evaporative condensers.
Source: From Shelton, B. G. et al., *Reducing Risks Associated with* Legionella *Bacteria in Building Water Systems*, Prevention and Control of Legionellosis (Technical Bulletin 2.4), PathCon Laboratories, Norcross, GA, 1995. With permission.

Level 1 is the review of routine maintenance programs as recommended by equipment manufacturers. Action Level 2 is a follow-up review for evidence of *Legionella* amplification. Action Level 3 represents low contamination, yet, elevated levels of concern. A review of the premises should be performed for direct and indirect bioaerosol contact with the aerosolized water and potential health risks associated with the contaminated water. Where outbreaks may become possible, Action Level 4 suggests cleaning and/or biocide treatment of the equipment. See Table 6.4 for specific remedial actions.

Helpful Hints

Legionella may be hidden and amplified within the cells of other microorganisms (e.g., protozoa). Thus, a negative result does not necessarily indicate the environmental source of a sample is free of *Legionella*. In such cases, the environmental professional should comment that low levels are an indication of low risk. There are no absolutes.

The occurrence of a potential misdiagnosis of disguised, elevated levels has resulted in some investigators choosing not to take samples. However, this choice may potentially result in not identifying those that are not misdiagnosed or disguised.

In a 1991 incident involving a Richmond, California, Social Security Office Building, two people died of Legionnaires' disease. The outbreak involved ten cases and created considerable negative publicity. On March 13, 1995, the U.S. Department of Justice settled the case rather than go to trial. The amount of the settlement was not disclosed, though it was apparently substantial. A question to ponder, "Does the policy not to test for Legionnaires' disease result in a reasonable standard of care?"[29]

Another incident involving exposures in a Jacuzzi on a cruise ship occurred in 1994. A more recent event involved misting of grocery produce and a link to

Table 6.4 Suggested Remedial Response Actions[24]

Action Level	Suggested Remedial Response
1	Review routine maintenance program recommended by the manufacturer of the equipment to ensure that the recommended program is being followed. The presence of barely detectable numbers of *Legionella* represents a low level of concern.
2	Implement Action 1, and conduct a follow-up analysis after a few weeks for evidence of further amplification. This level of *Legionella* represents little concern, but the number of organisms detected indicates that the system is a potential amplifier for *Legionella*.
3	Implement Action 2, and conduct a review of premises for direct and indirect bioaerosol contact with occupants and health risk status of people who may come in contact with the bioaerosols. Depending on the results of the premises review, action related to cleaning and/or biocide treatment of the equipment may be indicated. This level of *Legionella* represents a low but increased level of concern.
4	Implement Action 3. Then, cleaning and/or biocide treatment of the equipment is indicated. This level of *Legionella* represents a moderately high level of concern, since it is approaching levels that may cause outbreaks. It is uncommon for samples to contain numbers of *Legionella* at this level.
5	Immediate cleaning and/or biocide treatment of the equipment is clearly indicated. Conduct post treatment analysis to ensure effectiveness of the corrective action. The level of *Legionella* represents a high level of concern, since it poses the potential for causing an outbreak. It is uncommon for samples to contain numbers of *Legionella* at this level.

Source: From Shelton, B.G. et al., *Reducing Risks Associated with* Legionella *Bacteria in Building Water Systems*, Prevention and Control of Legionellosis (Technical Bulletin 2.4), PathCon Laboratories, Norcross, GA, 1995. With permission.

Legionnaires' disease. Each of these incidents may have been avoided had prior sampling been performed, elevated levels identified, and the sources remediated.

The criteria levels described above were developed by a laboratory that maintained a stringent quality assurance program, including in-house proficiency testing of the laboratory personnel to insure accuracy and reproducibility. The interpretative value of these data may not be applicable with quantitative values from other laboratories that do not have similar quality standards.

Other Pathogenic Bacteria

This section excludes *Legionella pneumophila*. These other "airborne pathogenic bacteria" of potential concern to the environmental professional are not as well studied yet still deserve attention.

Disease and Occurrence of Prominent Airborne Pathogenic Bacteria

As pathogenic bacteria may not be identified during air or diagnostic sampling, disease or the chance presence of pathogenic bacteria in a routine bioaerosol air sample may be the only means for suspecting its presence in a given environment. In alphabetical order, typical disease and occurrence of airborne pathogenic bacteria frequently posing a concern are herein discussed.

Bacillus anthracis[30,31]

The genera *Bacilli* are spore-forming, rod-shaped bacteria which require oxygen to grow. Although there are numerous species, only the *Bacillus anthracis* is known to be lethal to man. Many species have a commercial use in insect control in the agriculture and forest industry. Although *Bacillus anthracis* is the only deadly *Bacillus*, other forms may be opportunistic. Some species of *Bacillus* and several genus of *Clostridium* share the commonality of being the only pathogenic, spore-forming bacteria. However, *Bacillus anthracis* may cause disease through airborne transmission.

Mostly a disease of lower animals, *Bacillus anthracis* is transmissible to man via the skin, alimentary tract, and respiratory tract. Disease in man is typically an occupational disease associated with butchers, shepherds, herdsmen, and handlers of hides, hair, and fleece. During World War I, anthrax-contaminated articles (e.g., shaving brushes) from Asia and South America provided a source for infection. Today, most reports of anthrax disease are from third world countries and countries without venterinary public health programs (e.g., South and Central America, Southern and Eastern Europe, Asia, Africa, the Caribbean, and the Middle East).[32] The primary concern is for airborne exposures that result in pulmonary anthrax.

Pulmonary anthrax is due to inhalation of the microorganisms from the air. Although uncommon, it is the most dangerous. It occurs occupationally among those who handle/sort wool and fleece where the spores are floating in the air from the infected material. It is characterized by symptoms of pneumonia that frequently becomes fatal septicemia. The dose-response is low. Research indicates that only a few inhaled spores are required to produce disease. Immunization is possible, and infections are treated with antibiotics.

During World War II, *Bacillus anthracis* was researched heavily due to its airborne pathogenicity and resistance of the spores to drying. An effort was made to develop drug-resistant strains but it is not clear as to their success. They did, however, find that the spores can remain viable in soil for as long as sixty years.

Bacillus anthracis is also a hazard to textile workers working with imported animal hair. The primary concern may be airborne exposures, but the most likely source of infection is through the skin.

The genus *Bacilli* are Gram-positive rods, measuring 4.5 to 10 microns in length by 1 to 1.25 microns wide. They grow well on all common nutrient media, most rapidly between the temperatures of 41°C and 43°C in aerobic conditions. Spores can be destroyed by any of the following methods:

- Boil for ten minutes.
- Heat at 140°C for three hours under dry conditions.
- Disinfect with 3 percent hydrogen peroxide for one hour or with 4 percent potassium permanganate for fifteen minutes.

Corynebacterium diphtheriae[33]

Only one species of *Corynebacterium* is an airborne pathogen. That is *Corynebacterium diphtheriae* the cause of diphtheria.

Diphtheria results from airborne exposures to the bacterium. The *Corynebacterium diphtheriae* enters the body via the respiratory route, lodging in the throat and tonsils. Death may result from suffocation by blockage of the air passages and by tissue destruction from the toxin evolved.

Occasionally, it has affected the larynx, resulting in membranous croup, or the nasal passages, causing membranous rhinitis. Diphtheria infections of the conjunctiva and of the middle ear are less common, and cutaneous or wound diphtheria is only occasionally observed. However, wound diphtheria may be serious, resulting in a systemic infection. Systemic infections can severely affect the kidneys, heart, and nerves. Primary infection of the lungs and diphtheritic meningitis have been observed on rare occasion. Thus, most of the concerns for the diphtheria bacillus would be in hospital environments.

Corynebacterium diphtheriae are generally described as Gram-positive rods with irregular staining responses. The ends of each rod are swollen and respond greatest to staining, a characteristic unique to the genus. The bacteria measure 1 to 6 microns in length and 0.3 to 0.8 microns in width.

Mycobacterium tuberculosis[34,35]

Mycobacterium tuberculosis has three varieties (e.g., *hominis*, *bovis*, and *avian*). The *Mycobacterium tuberculosis* variety *hominis* is highly infectious via airborne transmission of the disease. Not only are there different varieties, but there are different strains.

Resistance to drug treatment occurs *in vitro* and *in vivo.* In the infectious stage, drug resistance is more common with the *Mycobacterium tuberculosis* than other pathogenic bacteria because of the extended, lengthy antibiotic treatments required to fight the disease. Their persistence allows greater chances for mutated strains to develop.

Tuberculosis is one of the most important communicable diseases in the world, affecting more than fifty million people. In the United States, over fifty thousand new cases are reported annually, and ten thousand resultant deaths occur per year. However, disease does not always develop in all those who have been exposed. There is a possible dose-response relationship.

Potential transmission occurs in public places and hospitals, typically by airborne droplets of contaminated sputum. Other modes of infection (e.g., genitourinary tract, conjunctiva of the eyes, skin, and alimentary tract) are less common.

Bacteria-laden droplets and dust particles are inhaled, settle in the lungs, and grow. In an individual with low resistance, an infection occurs. Extensive destruction and ulceration of the lung tissue progress to other parts of the body, predominately the spleen, liver, and kidneys. If not controlled by antibiotics, death may result.

The tuberculosis bacteria are acid-fast rods, measuring 2 to 4 microns in length by 0.3 to 1.5 microns wide. Specialized enriched media and aerobic conditions are required for growth and isolation of the bacillus. The optimum temperature for growth of the mammalian varieties is 37°C with a range of 30°C to 42°C.

Various Genera of *Pseudomonas*[36,37]

Pseudomonas includes over thirty species which are found in water, soil, and compost. Of the many species, there are only a few pathogenic types.

The most widely known pathogenic form is *Pseudomonas aeruginosa* (also referred to as *Pseudomonas pyocyanea*). It grows readily on all ordinary culture media and most rapidly between the temperatures of 30°C and 37°C. Although it typically requires oxygen, some will grow in anaerobic conditions. As *Pseudomonas aeruginosa* produces a bright blue-green color which diffuses into its substrate, a blue-green color has, at times, been observed on substrates (e.g., surgical dressings) where the bacteria grew.

Pseudomonas aeruginosa frequently occurs as a secondary invader of infected tissues or tissues which have been traumatized by operation. Children and immune-suppressed patients are particularly susceptible to infection. Oftentimes, infections are associated with the urinary and respiratory tracts. Abscesses may develop in different parts of the body, especially the middle ear. Although rare, it may also cause endocarditis, pneumonia, meningitis, and systemic pyocyanic infection. The latter generally occurs in patients with severe burn wounds where the bacterium enters and spreads throughout the body, causing endotoxic shock and focal necrosis of the skin and internal organs. As they do not respond well to antibiotics, systemic infections are almost always fatal. For this reason, the primary area of concern for this bacterium is in hospitals.

It should be noted that bacteriostatic detergents (e.g., cationic detergents) provide a growth environment for certain Gram-negative bacteria (e.g., *Pseudomonas aeruginosa*). Many of these soaps are used in skin antiseptics, mouthwashes, contact lens solutions, and disinfectants for hospitals.

Pseudomonas aeruginosa is also an insect pathogen and has consequently been considered for use as an insecticide. Its limited shelf life and potential to cause disease in humans has been a deterrent from its use as a pest control agent.

Pseudomonas pseudomallei infections may be asymptomatic or result in acute, toxic pneumonia or overwhelming septicemia. The organism has been isolated from moist soil, market fruits and vegetables, and well and surface waters in Southeast Asia. *Pseudomonas maltophilia* is suspect as well. However, over the past few years, the latter suspect has undergone numerous name changes and is now referred to as *Stenotrophomonal maltophili*. It is no longer classified as genus *Pseudomonas*.

The *Pseudomonas* bacteria are Gram-negative rods, measuring 1.5 to 3 microns in length by 0.5 microns wide. They grow well on all common nutrient media and grow most rapidly between the temperatures of 30°C and 37°C in aerobic conditions. Once airborne, the bacterium loses its viability with drying. So, exposures must occur shortly after the bacteria have become dispersed into the air.

Sampling and Analytical Methodologies

If pathogenic bacteria are suspect, the sampling equipment of choice for low to moderate levels (e.g., less than 10,000/m^3) are the culture plate impactors (e.g., Burkhard sampler and Andersen impactor). In situations involving potentially higher numbers, liquid impingers are preferred, primarily due to their ability to dilute the solution and plate from several diluents. Notwithstanding surface samples, filter air sampling is easier but may theoretically be used only for the spore-forming *Bacilli anthracus* for which analysis is not available commercially. *Mycobacterium tuberculosis* is also of questionable feasibility.

Although air sampling for *Mycobacterium tuberculosis* has been attempted by some environmental professionals, it is not recommended by most microbial laboratories. Where the concern involves this pathogen, alternative surface sampling may provide more reliable results. Otherwise, a negative air sample can not be considered conclusive as to the nonpresence of the pathogen.

In the laboratory, samples that are not already on a culture medium are plated. Filter surfaces are washed, the suspended material plated on nutrient media. Impinger solutions are also plated.

Most pathogenic bacteria will grow on tryptic soy agar or a nutrient blood agar, the same nutrients used to sample for total bacteria during indoor air quality studies. Although the nutrient media of choice is blood agar, the microbial laboratory chosen to perform the analysis should be consulted for their preferred media.

Incubation should take place at 35° C for most pathogenic bacteria. Pathogenic bacteria tend to grow best at elevated temperatures, and the higher temperatures restrict growth of most environmental bacteria that could overrun a sample.

With a few exceptions, colonies may be counted within one to two days. One such exception is *Mycobacterium tuberculosis*, which requires up to two weeks for colonies to become visible.

Where shipping is necessary, a count may take place upon receipt of a well-insulated shipment of culture plates. Impinger samples should be plated prior to shipping to avoid growth within the collection solution during transportation.

Interpretation of Results

Experience and careful consideration of comparative and/or diagnostic samples are important for developing sound conclusions. The sampling methodologies alone may damage some of the otherwise viable, unprotected bacteria. Thus, the mere identification of specific microbes in an indoor environment, particularly in hospitals, may be cause for remediation.

The outdoor environmental contribution of a pathogenic bacterium may also be important in evaluating the findings and devising a remediation plan. Where the levels are as high or higher outdoors, indoor exposure controls may be more difficult or, in some cases, not required.

Where air sampling is performed for *Mycobacterium tuberculosis*, positive findings of any level should be acted upon. However, negative results may not have any meaning. Negative findings should not be relied upon where false negatives, particularly involving drug-resistant strains, can lead to a lethal, false sense of security.

PATHOGENIC PROTOZOA[38,39]

Typically larger than the bacteria and mold spores, protozoa are unicellular microorganisms that are free-living and thrive in water. They may be located in damp soil, mud, drainage ditches, puddles, ponds, rivers, and oceans. Those that represent an occupational and environmental concern are the amoebae, measuring in size from 8 to 20 microns in diameter.

Amoeba move by flowing pseudopodia, or false feet, and usually live in fresh water. They are naked, or unprotected, during vegetation. Otherwise, they secrete a protective shell. They may be found in home humidifiers where they have been known to cause an allergic reaction called humidifier fever. The genera *Naegleria* and *Acanthanoeba* have been implicated in building-related illness, yet most amoeba-related disease is associated with waste treatment plants.

In the absence of water, the *Naegleria* can encyst to protect itself. These spherical cysts are 9 to 12 millimeters in diameter. The seriously parasitic amoebae, which are not typically found in fresh water, are not generally a concern.

Sampling and Analytical Methodology

Sampling equipment may be the Litton sampler, two in-series, all-glass impingers, or sieve plate impactors. High volume air samples should be collected close to the probable aerosolization source, avoiding other potential contaminant sources (e.g., not immediately under running water). Equipment should also be sterilized.

The Litton sampling tube should be cleaned, using 70 percent ethanol and submerged in two separate distilled water rinses for 30 to 60 minutes each. The last rinse should be collected and analyzed for background amoeba contamination of the tubing.

The all-glass impingers should be cleaned, and the final rinse water should be analyzed for background contamination. The final in-series samples (consisting of 150 milliliters of water) should be combined and analyzed collectively.

At the laboratory, each sample solution should be plated in five different amounts onto nutrient agar plates with the common bacteria *Escherichia coli* and incubated at between 43°C and 45°C. Pathogenic amoebae will grow more rapidly than non-pathogenic types in these conditions.

The amounts of solution per sample to be plated should be 0.01, 0.1, 1, 10, and 100 milliliters. They are treated differently. The 10-milliliter sample should be centrifuged at 500X gravity for 15 minutes, and the product plated. The 100 milliliter sample should be filtered through a 1.2 micron pore-sized, cellulose filter which must be quartered or halved and inverted in an agar plate. The smaller aliquots should be plated directly onto separate plates. Assure all plates have been properly labeled.

Incubation may take up to seven days at 45°C (or at 35°C for some of the less heat tolerant species), and then a series of more complex manipulation techniques are performed. The resultant suspension requires an additional three hours of time-lapsed examinations for concentration determination of thermophilic *Naegleria* species. Speciation requires an additional two weeks in *in vivo* mouse pathogenicity studies.

Interpretation of Results

Disease, coupled with reservoir source identification, is diagnostic and does not require quantitation. The existence of these heat-loving amoebae in a reservoir requires immediate remediation with an oxidizing biocide (e.g., sodium hypochlorite or hydrogen peroxide).

VIRUSES[40]

Viruses, the smallest living organisms, are obligate intracellular parasites. They are subdivided into animal, plant, and bacterial parasites. The size of animal viruses ranges from twenty to 300 nanometers.

Viruses are host specific and become invasive only when specific host organs become accessible. Host entry may be through any of a number of mechanisms, yet most enter through the respiratory tract.

A line of defense (e.g., nasal hairs and mucous secretions) must be passed. Then, the virus must be transported (e.g., through the blood supply) to its target cell preference(s). Once the target cells have been identified, the virus penetrates the cell wall barrier and takes command of the cell's replicating mechanism. The virus is reproduced within 6 to 48 hours. A protective structure is built around each replicated virus, and the progeny viron either destroys the host cell or forms a bud that allows it to pursue other host cells.

Animal viruses are usually recognized by the diseases they cause (e.g., AIDS). The greatest concerns with aerosolized animal viruses are influenza, measles, chicken pox, and some colds. Virulence is influenced by the following:

- Specific type of virus
- Concentration in an aerosol
- Aerosol particle size
- Individual susceptibility

Indoor contamination occurs in residences, offices, laboratories, hospitals, and animal confinement areas. Outdoor exposures occur around livestock, sewage treatment plants, caves, and water sources. Environmental factors affecting virus survivability are relative humidity, temperature, wind, ultraviolet radiation, season, and atmospheric pollutants.

Amplification of viruses does not occur without a host. Hence, increased numbers will not occur in water or organic substrates of air handlers. The air handlers will merely serve as a means of conveyance. Acids, extreme temperatures (e.g., less than minus 20°F), and drying may, however, damage or destroy most viruses. This information is important for an exposure limiting consideration for prevention and control of viruses and for handling samples.

Sampling is typically neither recommended nor requested. The presence of disease is generally tracked epidemiologically to the source or sources.

SUMMARY

Pathogenic microbes are rarely encountered in indoor air quality situations, and it is usually the result of an epidemic of cases when they become suspect. Sampling and analytical methodologies are difficult and time consuming. Some attempts are made in hospitals at proactive projections and occasional sampling. Yet, the investigator will generally be called in for assistance after an elevated number of suspect disease cases are associated with a building. The challenge may be more that of confirmation of a suspect agent and locating the source.

REFERENCES

1. Bailey, M. Robert, Ph.D. and E. Scott. *Diagnostic Microbiology*. C.V. Mosby Company, St. Louis, Missouri, 3rd Edition, 1970.
2. Conant, Norman F., Ph.D., et. al. *Manual of Clinical Mycology*. W.B. Saunders Company, Philadelphia, 3rd Edition, 1958.
3. U.S. Department of Health, Education, and Welfare. *Occupational Diseases: A Guide to Their Recognition*. U.S. Government Printing Office, Washington, D.C., Revised Edition, June 1977.
4. Rhame, Frank S., M.D. Endemic Nosocomial Filamentous Fungal Disease: A Proposed Structure for Conceptualizing and Studying the Environmental Hazard. *Infection Control*. 7(2):126 (1986).
5. Krasinski, Keith, M.D., et. al. Nosocomial Fungal Infection During Hospital Renovation. *Infection Control*. 6(7):278-82 (1985).
6. Weems, J. John, M.D., et. al. Construction Activity: An Independent Risk Factor for Invasive Aspergillosis and Zygomycosis in Patients with Hematologic Malignancy. *Infection Control*. 8(2):71-5 (1987).
7. Cross, Alan S., M.D. Nosocomial Aspergillosis: An increasing problem. *Journal of Nosocomial Infection*. 4(2):6-9 (1985).
8. Mascola, J.R. and L. Rickman. Infectious causes of carpal tunnel syndrome: case report and review. *Rev. Infect. Disease*. 13(5):911-7 (1991).
9. Kilburn, C.D. and D. McKinsey. Recurrent massive pleural effusion due to pleural, pericardial, and epicardial fibrosis in histoplasmosis. *Chest*. 100(6):1715-7 (1991).
10 Center for Disease Control. Update on Coccidioidomycosis in California. *Morbidity and Mortality Weekly Report*. 43(23):421-3 (1994).
11. Chen, G.H., et. al. Case Records of the Massachusetts General Hospital—Weekly Clinicopathological Exerciser. *New England Journal of Medicine*.

330(7):490-6 (1994).
12. Paya, C.V. Fungal Infections in solid-organ transplantation. *Clinical Infectious Diseases.* 16(5):677-88 (1993).
13. Pfeiffer, T.J. and D. Ellis. Environmental isolation of Cryptococcus neoformans var. gattii from Eucalyptus tereticornis. *Journal of Medical and Veterinary Mycology.* 30(5):407-8 (1992).
14. Morris, George K., Ph.D. and Brian G. Shelton M.P.H. *A Suggested Air Sampling Strategy for Microorganisms in Office Settings.* [Technical bulletin] PathCon Laboratories, Norcross, Georgia, 1994.
15. ACGIH Committee on Bioaerosols. *Guidelines for the Assessment of Bioaerosols in the Indoor Air Environment—Fungi.* ACGIH, Cincinnati, Ohio, 1989.
16. Gallup, Janet and Mirian Velesco, Dr.P.H. *Characteristics of Some Commonly Encountered Fungal Genera.* Environmental Microbiology Laboratory, Inc., Daly City, California, 1999.
17. Ajello, Libero, Ph.D., et. al. *Microbes in the Indoor Environment: A Manual for the Indoor Air Quality Field Investigator.* Path Con Laboratories, Norcross, Georgia, 1998.
18. Department of Pathology, University of Texas Medical Branch, Medical Mycology Research Center [last entry May 2000]. Available on the World Wide Web: (http://fungus.utmb.edu/mycology/thefungi.html)
19. Burge, Harriet A. *Bioaerosols.* Lewis Publishers, Boca Raton, Florida, 1995. p. 259.
20 Stewart, F.S. *Bacteriology and Immunology for Students of Medicine.* Williams and Wilkins Company, Baltimore, Maryland, 9th Edition, 1968.
21. Morris, George K., Ph.D. and B. Shelton, M.P.H. *Legionella in Environmental Samples: Hazard Analysis and Suggested Remedial Actions.* [Technical Bulletin 1.4] PathCon Laboratories, Norcross, Georgia, 1995.
22. Brock, Thomas D. and M. Madigan. *Biology of Microorganisms.* Prentice Hall, Englewood Cliffs, New Jersey, 6th Edition, 1991. pp. 518-520.
23. Burge, Harriet A. *Bioaerosols*—Legionella *Ecology.* Lewis Publishers, Boca Raton, Florida, 1995. pp. 49-76.
24. Shelton, Brian G., et. al. *Reducing Risks Associated with* Legionella *Bacteria in Building Water Systems.* Prevention and Control of Legionellosis. [Technical Bulletin 2.4] PathCon Laboratories, Norcross, Georgia, 1995.
25. *Legionella* from misting in grocery store. *New York Times,* January 11. 139:A1(N) 1990.
26. Hlady, W. Gary, et. al. Outbreak of Legionnaire's disease linked to a decorative fountain by molecular epidemiology. *American Journal of Epidemiology.* 138(8):555-62 (1993).
27. Alcamo, I. Edward, Ph.D. *Fundamentals of Microbiology.* 3rd Edition, Benjamin/Cummings Publishing Company, Inc., Redwood City, California. p. 243.

28. Gorman, George W., J. Barbaree, J. Feeley, et. al. *Procedures for the Recovery of* Legionella *from the Environment.* [Bulletin] CDC, Atlanta, Georgia, November 1992.
29. Shelton, Brian G. Social Security Building incident where settlement is alledged to have been quite expensive. [Oral communication] PathCon Laboratory, Norcross, Georgia, July 1995.
30 Atlas, Ronald M. and R. Bartha. *Microbial Ecology: Fundamentals and Applications.* Benjamin/Cummings Publishing Company, Menlo Park, California, 1987. p. 476.
31. Burrows, William, Ph.D. *Textbook of Microbiology.* W.B. Saunders Company, Philadelphia, 1968. pp. 614-619.
32. Center for Disease Control. Anthrax—General Information. [last entry April 4, 2001]. Available on the World Wide Web: (www.cdc.gov/ncidod/dbmd/diseaseinfo/anthrax_g.html)
33. Brock, Thomas D. and M. Madigan. *Biology of Microorganisms.* Prentice Hall, Englewood Cliffs, New Jersey, 6th Edition, 1991. pp. 513-514.
34. ACGIH Committee on Bioaerosols. *Guidelines for the Assessment of Bioaerosols in the Indoor Air Environment—Bacteria.* ACGIH, Cincinnati, Ohio, 1989.
35. Stewart, F.S. *Bacteriology and Immunology for Students of Medicine.* Williams and Wilkins Company, Baltimore, Maryland, 9th Edition, 1968. pp. 357-60.
36. Ibid. pp. 278-280
37. Ibid. pp. 282-283.
38. ACGIH Committee on Bioaerosols. *Guidelines for the Assessment of Bioaerosols in the Indoor Air Environment—Protozoa.* ACGIH, Cincinnati, Ohio, 1989.
39. Burge, Harriet A. *Bioaerosols.* Lewis Publishers, Boca Raton, Florida, 1995. pp. 122-123.
40. ACGIH Committee on Bioaerosols. *Guidelines for the Assessment of Bioaerosols in the Indoor Air Environment. Viruses.* ACGIH, Cincinnati, Ohio, 1989.

Chapter 7

TOXIGENIC MICROBES

Toxigenic microbes are mycotoxins and endotoxins, the toxins associated with fungi and bacteria, respectively. As toxigenic microbes have often been overlooked, environmental professionals are becoming more aware of their potential impact on the indoor air quality.

Organic dust toxic syndrome is thought to be associated with a complex mix of mycotoxins, endotoxins, antigens (allergenic material), and glucans. This is a poorly characterized condition similar to humidifier fever. Flu-like symptoms include fever, chills, muscle aches, malaise, and chest tightness, and illness is short lived, usually less than 24 hours. Exposures and symptoms are generally associated with high dust levels in composting and some agricultural operations.

Not only do mycotoxins and endotoxins exposures result in adverse health effects, but they may have a positive impact as well. Whereas one mold may cause cancer, another may protect against pathogens. Excessive exposures to endotoxins may result respiratory distress, shock, and death, but low level exposures may stimulate the immune system, which may help reduce the risk of cancer.

Each of the above-mentioned toxigenic microbes is worthy of mention and should not be overlooked when assessing an indoor air environment. They are discussed herein.

MYCOTOXINS[1,2]

Mycotoxins are toxic by-products of the metabolic process of fungi. Those that produce the mycotoxins gain a competitive edge over other microorganisms for food. Many fungi produce at lease one mycotoxin. Some produce many different mycotoxins, and others do not produce any. See Table 7.1.

Exposues to mycotoxins may occur by eating poisonous/hallucinogenic mushrooms, eating contaminated foods, breathing heavily contaminated air, or handling contaminated materials with unprotected hands. The greatest impact has been on agriculture and livestock. The feeds become contaminated with toxin-producing fungi. The livestock eat the heavily contaminated feeds with a possible outcome of death.

The by-products have been identified as the fungal mycelium, spores, and food substrates. For instance, *Aspergillus fumigatus* produces an aflatoxin in the spore

Table 7.1 Mycotoxins Associated with Certain Molds[3]

Mold	Mycotoxin
Acremonium spp.	Cephalosporin (antibiotic)
Alternaria alternata and *Phoma sorghina*	Tenuazoic acid
Aspergillus clavatus	Cytochalasin E Patulin
Aspergillus flavus and Aspergillus parasiticus	Aflatoxins
Aspergillus fumigatus	Fumitremorgens Gliotoxin
Aspergillus nidulans, Aspergillus versicolor and Cochliobolus sativus	Sterigmatocystin
Aspergillus ochraceus, Penicillium verrucosum and *Penicillium viridicatum*	Ochratoxin A
Cladosporium spp.	Epicladosporic acid
Cladosporium cladosporioides	Cladosporin & emodin
Claviceps purpurea	Ergot alkaloids
Fusarium graminearum	Deoxynivalenol Zearalenone
Fusarium moniliforme	Fumonisins
Fusarium moniliforme	Fumonisins
Fusarium pose and *Fusarium sporotrichoides*	T-2 toxin
Penicillium chrysogenum	Penicillin (antibiotic)
Penicillium crustosum	Penitrem A Roquefortine C
Penicillium expansum	Citrinin Patulin Roquefortine C
Penicillium griseofulvum and *Penicillium viridicatum*	Griseofulvin
Pithomyces chartarum	Sporidesmin Phylloerythrin
Stachybotrys chartarum (*atra*)	Satratoxins Verrucarins Roridins Stachybocins
Tolypocladium inflatum	Cyclosporin

wall. This aflatoxin rapidly may diffuse out of the wall and into a water reservoir. However, mycotoxins are not known to become airborne without a substrate (e.g., mold spores and components).

Possible short-term symptoms of excessive exposures to mycotoxins include nausea, vomiting, dermatitis, cold and flu symptoms, sore throat, fatigue, and diarrhea. Some other reported effects that are more species dependent are eczema, photosensitization, hemolysis, hemorrhage, and impaired or altered immune function that can result in opportunistic infections. Long-term effects are even more species dependent. Specific species may be hepatotoxic (e.g., *Aspergillus flavus*), cytotoxic (e.g., *Aspergillus fumigatus*), teratogenic (e.g., *Penicillium viridicatum*), tremogenic (e.g., *Penicillium expansum*), tumorigenic (e.g., *Aspergillus fumigatus*), nephrotoxic (e.g., *Aspergillus clavatus*), mutagenic (*Aspergillus parasiticus*), and/or carcinogenic (e.g., *Cochiobolus sativus*).[4]

Some mycotoxins are beneficial to man in the form of antibiotics (e.g., *Penicillium chrysogenum*). The antibiotics are toxic to pathogenic bacteria. The antibiotic mycotoxins revolutioned the world and have saved many lives. Many mycotoxin-producing fungi are used in making cheeses (e.g., *Penicillium expansum* makes Roquefort cheese).

There is a dose-response relationship for mycotoxins similar to that of chemical toxins. The impact of mycotoxins is based upon the type of toxin, the animal species impacted, the route of entry, and susceptibility of the individual.[5]

Disease and Occurrence

Throughout history, mycotoxins have played a significant role in the health of man and animal. Toxicity and dramatic exposures have typically been associated with ingestion of contaminated foods. Ergotism, also known as Saint Anthony's Fire, was documented as early as 430 B.C. by the Spartans. It was caused by ingestion of rye that was contaminated with alkaloid containing fungi. Symptoms were gangrene, convulsions, and death.

In 1960, feed (i.e., Brazilian peanut meal) that was contaminated with *Aspergillus flavus* killed 100,000 turkeys in the United Kingdom. The toxins that were identified came to be named aflatoxins.

Aspergillus flavis and *Aspergillus parasiticus* create aflatoxins that are carcinogenic and may cause some short-term health effects which have yet to be clarified by medical researcher. These molds grow predominately in warm, humid environments, and are typically associated with peanuts, grains, sweet potatoes, corn, peas, and rice. Aflatoxins have been reported to cause liver carcinomas in animals and are believed to be amongst the most potent of carcinogens in man.[6] They are primarily associated with liver cancer, but there have been some concerns raised regarding inhalation and lung cancer.

There is one report of two deaths due to pulmonary adenomatosis thought to be related to airborne aflatoxin exposures. Although autopsies indicated the presence of aflatoxin in both patients' lungs, the findings were not conclusive.[7] The verdict regarding airborne aflatoxins is still out.

Trichothecene toxins are a by-product of the metabolism of *Fusarium*, *Acremonium*, *Trichoderma*, *Myrothescium*, and *Stachybotrys*. Reported symptoms include headaches, sore throats, hair loss, flu-like illness, diarrhea, fatigue, dermatitis, and generalized malaise.

The Year 2000 saw the birth of a new cottage industry—the quest for and remediation of toxigenic molds. Based on a multimillion-dollar lawsuit, *Stachybotrys chartarum* was tagged by the news media as the "death mold." Investigators began targeting *Stachybotrys chartarum* with fervent interest in an effort to stem the tide of panic. Some "experts" began targeting the genus *Penicillium* as well. Indoor environments with water-damaged building materials and potentially toxigenic fungi were posted with "Biohazard" signs and remediated at great expense to the building owner or insurance company. Most, if not all, of the investigations mentioned above involving expensive remediation have yet to confirm the presence of mycotoxins. So, as of the publication of this book, the saga continues.

Some species of *Penicillium* produce mycotoxins. Some species (e.g., *Penicillium viridicatum*) produce mycotoxins that are potentially tumorigenic, teratogenic, and hepatotoxic.[4] Others produce antibiotics (e.g., *Penicillium chrysogenum*).

To complicate matters, toxins are not always produced. The nutrient (or substrate) upon which the mold grows impacts its metabolites. Indoor environments and laboratory nutrients may result in a variation from the norm. Thus, the presence of amplified numbers of mycotoxin-producing fungi in a building does not necessarily mean that it is producing the toxin. Likewise, the failure of a mold to produce toxins in laboratory conditions does not necessarily mean that it does not produce the toxin in outdoor environments.

Other factors that may affect production of mycotoxins are pH, temperature, and water. For instance, aflatoxin production is associated with drought and heat stress.

A recent research publication indicated:

> *In many cases, the presence of fungi thought to produce the mycotoxins was not correlated with the presence of the expected compounds. However, when mycotoxins were found, some toxigenic fungi usually were present, even if the species originally responsible for producing the mycotoxin was not isolated. We concluded that the identification and enumeration of fungal species present in bulk materials are important to verify the severity of mold damage but that chemical analyses are necessary if the goal is to establish the presence of mycotoxins in moldy materials.*[8]

In other words, the presence of a toxin-producing mold does not always a toxin make.

In summary, most cases of mycotoxin poisoning occur in rural and agricultural settings as a result of ingestion and/or skin contact. There have been minimal studies and research into the possible relationship between airborne mycotoxins and its impact on humans.

Sampling and Analytical Methodologies[2]

There are two different approaches to sampling for mycotoxins. The fungi may be identified with an emphasis on fungi identification for those molds that may potentially create toxic by-products, or the investigator may focus on toxins identification. The latter, although preferred, is neither easy nor technically feasible for commercial laboratories.

Fungi Identification

Species identification is a must and can only be determined through viable air sampling. Yet, where the fungi are to be identified from culturing, many of the toxin-producing fungi may not compete well with the other fungi, or they may no longer be viable where the toxins still persist in nonviable fungi. Special nutrient media may be necessary depending on the investigator's emphasis (e.g., *Stachybotrys*), the laboratory, and the laboratory methodologies(e.g., special cellulose agar).

Mycotoxins have the same effect—alive or dead—and the effect of toxins is based on a dose-response relationship. For this reason, total viable/nonviable air sampling is as important as viable air sampling. Both types of sampling should be performed in tandem. The methods are described in Chapters 5 and 7.

Toxin Identification

Sampling methodologies for mycotoxins have recently been developed and have become available commercially. Methods include air, contaminated-surface wipes, surface dust, and bulk sampling.

Air sampling requires a sufficient amount of air volume to enable capture of greater than 100,000 spores per sample. Based on an airborne mold spore exposure of 100 counts/m^3, the latter 100,000 spore samples would require an air volume of not less than 10^6 liters, or 46 days of continuous sample at 15 liters per minute. If the airborne mold spore exposure were 1,000 counts/m^3 (an unlikely indoor airborne mold spore count), the sample duration required would be about 5 days. Prior to taking mycotoxin air samples, airborne mold spore levels should be determined using one of the viable/nonviable sampling methodologies in Chapter 5. With known airborne spore counts, the investigator can calculate minimum air volume required to collect 100,000 spores. Report sample number (or location) and air volume to the laboratory.

Contaminate-surface wipe sampling may only be performed of areas where molds have been identified, and surfaces are heavily contaminated. A dusty surface is not sufficient and requires a different sample technique. Sample supplies should include sterile methanol wipes and/or swabs. Some laboratories supply field kits that consist of enclosed cotton swabs with methanol as well as templates to delineate the area(s) sampled. Lightly contaminated surfaces may require larger surface area sampling than heavily contaminated surfaces, and the size of the area template or number of areas sampled for one analytical run should be adjusted accordingly. Report sample number (or location) and surface area sampled to the laboratory.

Surface dust samples require the collection of a minimum of one teaspoon of dust. For general dust sampling methods, see the "Surface Dust Collection" section in Chapter 15. The surface dust sampling methods for micotoxins should be performed with supplies that have been cleaned (e.g., wash scrapping items and collection attachments with soapy water, rinse, and thoroughly dry). A minimum area of 1 square foot should be sampled. Report sample number (or location) and surface area sampled to the laboratory.

Bulk samples may be taken of carpeting, insulation, gypsum board, wood flooring, sheet vinyl wallpaper, and/or other materials that have visible evidence of mold growing on the surface(s). The size of the material selected to sample should be at least 1 inch square for heavy contamination and as much as 1 square foot for light contamination. Using clean tools, cut the material out, place it in a zip-lock bag, and label the bag with the sample number and/or location.

All samples should be labeled at the time each sample is taken, and the investigator must report to the laboratory sample number (or location) and air volume or surface area sampled. A log be maintained by the investigator of sample number(s) with a detailed description of sample sites (e.g., exact location where the sample was taken, type of material, and appearance of the sample), environmental conditions, and other information that may be applicable regarding the sample. All samples with required laboratory data and chain-of-custody should be shipped in a chilled container (e.g., Styrofoam cooler with a cool pack).

Analysis involves dust extraction and high-pressure liquid chromatography or thin-layer chromatography. See Table 7.2 for a sample of laboratory quantification results of mycotoxins produced by some cultivated fungal species.

In response to food testing required by the U.S. Food and Drug Administration (FDA), owners of granaries and corn storage facilities routinely perform a quick colorimetric test for the presence of aflatoxin. Although the FDA probes larger samples and has analyses performed by a commercial laboratory, colorimetric test kits (based on immunoassay tagging of the mycotoxin) are available commercially and are readily used for screening by the storage facilities.[9] This approach has yet to be introduced and/or used by environmental professionals to screen air samples.

Another technique that may hold some promise for environmental professionals is the ELISA detection kits. They are presently used to screen for mycotoxins in grains, nuts, and spices. The kits are commercially available and can be performed by the investigator within two hours. They are qualitative and quantitative in separate kits for the following:

- Aflatoxin
- Fumonisin
- Ochratoxin
- T-2 Toxin
- Vomitoxin
- Zearalenone

Table 7.2 Quantified Mycotoxins Produced by Cultivated Mold Species[10]

Mold	Toxin	Spores (ng/g)
Aspergillus fumigatus	F migaclavin C	930,000
Aspergillus niger	Aurasperone	460,000
Aspergillus parasiticus	Aflatoxin B1	16,600
Penicillium oxalicum	Secalonic acid D	1,890
Aspergillus parasiticus	Norsolorinic acid	280

Interpretation of Results

There are no definitive guidelines for interpretation of results. First, the fungi able to produce toxins may not produce these mycotoxins under the conditions that the contaminant is being assessed. Second, if the toxins could be identified, there are no known comparative values by which to determine that a problem does indeed exist.

On the other hand, a thorough assessment of dust for the presence of mycotoxins in a given environment may permit an investigator to determine if the suspect mycotoxins are present or not. The type of toxin sought should be based on fungal species found. Finding that the suspect mycotoxins present could be important information in itself, or outdoor samples may be compared to indoor samples.

Also, tell-tale symptoms may be used in conjunction with the presence of fungi known to potentially cause the same symptoms. Although this technique has been used more than the mycotoxin identification method, the symptoms must be properly diagnosed and some symptoms may be caused by other environmental exposures. In the latter case, a source misdiagnosis could result in misdirected efforts to remediate. If a fungus (e.g., *Penicillium*) is remediated and the problem is associated with a toxic chemical (e.g., Chlordane), the true cause of the symptoms remain unidentified.

The environmental professional is on his own or must rely on guidance from one with experience in this area of expertise. The laboratory chosen to perform the work should be able to either provide guidance or direct you to someone competent in that area. Each case will be unique and depend on several variables including intent and sampling strategy. All factors must be considered in the final interpretation.

BACTERIAL ENDOTOXINS[11,12,13]

Endotoxins are toxic components of the lipopolysaccharide, cellular membrane of Gram-negative bacteria. Although the bacteria may be rendered nonviable, the endotoxin retains its toxicity, even through extremely high temperatures (up to 110°C) that may not normally be exceeded in an autoclave.[11] Their impact on the individual is dose related, and airborne levels have been reported from a variety of work environments.

The most commonly reported cases have involved processing of vegetable fibers, fecal material in agriculture, and human waste treatment plants. Most worker exposures occur in cotton gins/mills, swine confinement buildings, poultry houses, grain storage facilities, sewage treatment facilities, and wood chip processing/saw mills. Other areas include the aerosolizing of machining fluids in metal processing,[14] aerosolized contaminated water supplies in hospitals,[15] and humidified office buildings.[16] See Table 7.3 for a recap and amplification on the environmental settings in which endotoxins have been identified at levels significant enough to cause illness.

Symptoms typically involve elevated temperatures. In the cotton industry, this is referred to as mill fever. Onset of fever is followed by malaise, respiratory distress (e.g., coughing, shortness of breath, and acute air flow obstruction, diarrhea, vomiting, hemorrhagic shock, tissue necrosis, and death.[17] However, the latter, more serious symptoms are rare.

Some researchers have also been able to demonstrate acute changes in the respiratory FEV_1 and suggest that repeated exposures may cause a syndrome similar to chronic bronchitis.[18] Other investigators feel that endotoxins add to the virulence of a parasite, enhancing the perils of disease. Once again, hospitals provide a special niche for such occurrences. Paradoxically, on the same note, it is also thought that the endotoxins that threaten one's health can enhance the body's immune system by challenging pathogenic bacteria, viral infections, and cancer.[19] While increasing the risk of hypersensitivity disease, endotoxins are a powerful, nonspecific stimulant to the immune system.[20]

Sampling and Analytical Methodologies

Bulk sampling of humidifier reservoirs and other potential endotoxin sources is considered the most reliable means of sampling for endotoxins. All samples should be collected in sterilized glassware and analyzed promptly, before microbial and subsequent endotoxin amplification can occur. Sterilization of all glassware and associated laboratory equipment should be baked at 210°C for one hour to destroy most pre-existing endotoxin contaminants. Plastic items should also have been sterilized (e.g., ethylene oxide) or sonicated in endotoxin-free one percent triethylamine. As they may adsorb large amounts of lipopolysaccharides, supplies made of polypropylene plastics should be avoided. Polystyrene is preferred.

Air sampling has been performed with limited success. Attempts have been made by dust/particle collection techniques. The methods of collection have included total dust samplers, cascade impactors (size separators), vertical elutriators (used for respirable dust sampling in the cotton industry), cyclones (used for personal respirable dust sampling), and midget impingers (personal samples which collect particles greater than 1 micron in size). As endotoxins act primarily in the lower region of the lungs (which mostly involves particles of less than 5 microns), respirable dust samplers are preferred. For those considering the possibility of viable sampling for bacteria to get a gauge as to endotoxin levels, endotoxin and viable bacteria samples have not been found to correlate.[21]

Table 7.3 Occupational Environments where Endotoxins Have Been Identified and Known to Pose a Problem[16]

Cotton Gins/Mills
Swine Confinements
Grain Storage, Handling, and Processing Facilities
Poultry Barns
Sewage Treatment and Processing Facilities
Wood Chipping Operations
Saw Mills
Flax Mills
Machine Shops
Fiberglass Manufacturing[19]

Sample collection flow rates have varied, based upon the type of sampler. They have been as low as 2 or 4.5 liters per minute for total dust to as high as 7.4 liters per minute with the vertical elutriator. Although not specified, one study suggests that in order to detect down to 14 picograms per cubic meter, a minimum of 7.5 cubic meters of air should be sampled.[22] Another study sampled at 2 liters per minute for the duration of a workshift, for a volume of 960 liters.[23] It is quite apparent that laboratories vary in their analytical methods and, therefore, their detection capabilities. For this reason, the laboratory should be consulted prior to sampling.

Filter media are also worthy of mention. In one study, recovery of endotoxin from four different types of filters was compared to a no filter control. The recovery rates, as reflected in Table 7.4, were extremely poor (e.g., polyvinyl chloride membrane filter) to moderate (e.g., mixed cellulose ester membrane filter).[21] In a later publication, the same investigator who performed the filter media testing proposed the use of 0.4-μm polycarbonate filters in a standard protocol for sampling endotoxins.[24]

Upon selection of sampling media, all supplies should be pre-cleaned and sterilized, including the filters and cassettes. Cassettes can be cleaned by sonication in endotoxin-free 1 percent triethylamine, and all other supplies that might come in contact with the sample (including the laboratory implements) should be autoclaved at 210° C for 1 hour.[23] As the ability to perform these processes is generally not within the realm of the environmental professional, supplies (including filters) are best provided by the laboratory.

After collection, the samples and a field blank (unsampled filter) should be sealed in airtight plastic enclosures, placed in a cold storage container, and sent immediately to the laboratory. Do not freeze the samples.[22] Each of the samples should be analyzed at the same time, by the same method, within the same laboratory.

In the 1940s, endotoxin was measured by injecting rabbits and monitoring increased body temperatures. The sensitivity was 100 picograms, but rabbits and preparations were inconsistent.[23] Referred to as the Pyrogen Test, this method remained in place until the discovery of the unique effects endotoxin had on the blood cells of the horseshoe crab.

In 1977, the U.S. Food and Drug Administration licensed the more sensitive Limulus Amebocyte Lysate (LAL) Assay as an alternative to the Pyrogen Test. An extract is made from amebocytes of the *Limulus polyphemus* (or horseshoe crab). In the presence of endotoxins, clotting occurs. There is a relationship between the amount of endotoxin present and the rate/amount of clotting.

Samples are extracted and tested according to one of four principal methodologies that are based upon blood cell clotting. Extraction methods vary as much as the analytical methods. In a proposed standard method, however, the recommended procedure is to extract samples in 0.05 M potassium phosphate, 0.01 percent triethylamine, pH 7.5, using bath sonication.

Analytical methods include the gel-clot test, calorimetric test, chromogenic test, and Kinetic-Turbidimetric Limulus Assay with Resistant-parallel-line Estimates (KLARE) Test. Due to its precision, sensitivity, resistance to interferences, internal validation of estimates, and ability to provide quantitative as well as qualitative information, the KLARE test is preferred by researchers and there is an attempt to standardize the use of this test as well.[25] However, despite all the studies and research, the choice of method is going to be laboratory dependent.

Interpretation of Results

Results are reported in terms of equivalent weight per volume (e.g., ng/ml or ng/m^3), equivalent mass (e.g., ng/mg) or potency endotoxin units per volume (e.g., EU/m^3).

Table 7.4 Recovery Efficiencies of Endotoxin from Filter Media

Filter Type	Recovery (ng of NP-1 activity/ ng LPS added)	Percent Recovered as Compared to the Controls (%)
Polyvinyl chloride (5.0-μm pore membrane)	0.064	10.8
Polyflon (woven Teflon)	0.206	34.4
Teflon membrane (1.0-μm pore membrane)	0.229	38.2
Cellulose mixed ester (0.45-μm pore membrane)	0.252	42.1
Lipopolysaccharide (LPS) control (no filter)	0.599	

NP-1 = Reference endotoxin.
Excerpted from *Endotoxin Measurement: Aerosol Sampling and Application of a New Limulus Method.*[23]

An endotoxin unit is defined as the potency of 0.10 ng of a reference standard endotoxin.[24] The U.S. Reference Standard EC-5 for endotoxin has a conversion factor of 10 EU per nanagram of substance.

With the results in hand, interpretation is precarious at best. There have been no standardized means in LAL preparations and interpreting endotoxin results. Furthermore, variation among laborators have been as much as 1,000 fold. However, results are less than a 10 fold difference where a laboratory uses the LAL from the same lot.[26] It, thus, becomes evident that a single exposure limit is not feasible, but relative limit values for samples analyzed at the same time, under similar methodologies and LAL lot, are indicated.

Many of the more prominent researchers and commercial laboratories tend toward relative limit values. This involves comparative sampling between complaint and non-complaint areas. When sampling for endotoxins in a metal fabrication shop around the cutting fluids, a comparative sample may be collected in an office. When sampling in a turkey processing plant, a comparative sample may be collected in the administrative area. Relative differences should shift simultaneously along with variances created by the different approaches.

According to the ACGIH Committee on Bioaerosols, endotoxin levels between 10 and 100 times the background levels (e.g., noncomplaint areas), coupled with health effects consistent with elevated exposures, should be remediated. Thus, 10 times the background levels is the ACGIH proposed relative limit value.[26]

A safety factor should also be considered. The ACGIH suggests 30 times the background levels, in the absence of symptoms.[26]

SUMMARY

Mycotoxins are toxins produced on the surface molds—alive or dead. They are a highly controversial topic brought on by the media, and they are here to stay. Often, they are implicated by mold types (e.g., *Stachybotrys chartarum*), but they are not always produced. So, the assumption that the toxigenic mold is producing mycotoxins is made. There are now methods available to confirm or deny the presence of many of the mycotoxins. The investigator would be well advised to confirm the presence of mycotoxins prior to making costly decisions.

Endotoxins are toxins produced by Gram-negative bacteria. Although they produce some of the symptoms occasionally encountered in indoor air quality investigation, they are frequently overlooked as a potential source.

Both mold and bacterial toxins are infrequently assessed, and it is difficult to trace down a laboratory with the capabilities to do the analyses. Keep in mind, over 50 percent of the investigations in one study group remain unsolved as to the source. The investigator should be aware and consider the possibility of microbial toxins when assessing indoor air quality.

REFERENCES

1. Pleil, J.D. Demonstration of a Valveless Injection System for Whole Air Analysis of Polar VOCs. *Proc. 1991 Int. Symp. Measurement of Toxic and Related Air Pollutants.* Air and Waste Management Association, Pittsburgh, Pennsylvania, 1991.
2. ACGIH Committee on Bioaerosols. *Guidelines for the Assessment of Bioaerosols in the Indoor Air Environment—Mycotoxins.* ACGIH, Cincinnati, Ohio, 1989.
3. Sheldon, Brian G. Social Security Building incident where settlement is alleged to have been quite expensive. [Oral communication] PathCon Laboratory, Norcross, Georgia, July 1995.
4. ACGIH. *Bioaerosols: Assessment and Control.* ACGIH, Cincinnati, Ohio, 1999. p. 24-3.
5. Cox, Christopher S. and C. Wathes. Bioaerosols Handbook. CRC/Lewis Publishers, Boca Raton, Florida, 1995. p.375.
6. Burge, Harriet A. *Bioaerosols.* Lewis Publishers, Boca Raton, Florida, 1995. p. 90.
7. Cox, Christopher S. and C. Wathes. *Bioaerosols Handbook.* CRC/Lewis Publishers, Boca Raton, Florida, 1995. pp. 375-6.
8. Tuomi, T, et. al. Applied Environmental Microbiology. 66(2), 2000. pp. 1899-1904.
9. Aflatoxin–Nature's most potent carcinogen. [Handout] Neogen Corporation, Lansing, Michigan, 1996.

10. ACGIH. *Bioaerosols: Assessment and Control.* ACGIH, Cincinnati, Ohio, 1999. p. 24-4.
11. ACGIH Committee on Bioaerosols. *Guidelines for the Assessment of Bioaerosols in the Indoor Air Environment—Endotoxins.* ACGIH, Cincinnati, Ohio, 1989.
12. Burge, Harriet A. *Bioaerosols.* Lewis Publishers, Boca Raton, Florida, 1995. p. 78.
13. ACGIH. *Bioaerosols: Assessment and Control.* ACGIH, Cincinnati, Ohio, 1999.
14. Gordon, Terry. Acute Respiratory Effects of Endotoxin Contaminated Machining Fluid Aerosols in Guinea Pigs. *Fundamental and Applied Toxicology.* 19:117-23 (1992).
15. Burrell, Robert. Human Responses to Bacterial Endotoxin. *Circulatory Shock.* 43:137-53 (1994).
16. Jacobs, Robert R. Airborne Endotoxins: An Association with Occupational Lung Disease. *Applied Industrial Hygiene.* 4(2):50 (1989).
17. ACGIH. *Bioaerosols: Assessment and Control.* ACGIH, Cincinnati, Ohio, 1999. p. 23-2.
18. Reynolds, Stephen J. and D. Milton. Comparison of Methods for Analysis of Airborne Endotoxin. *Applied Occupational Environmental Hygiene.* 8(9):761-7.
19. Jacobs, Robert R. Airborne Endotoxins: An Association with Occupational Lung Disease. *Applied Industrial Hygiene.* 4(2):52 (1989).
20. Rietschel, Ernst Theodor, and Helmut Brade, *Scientific American.* 267(2):55-61 (1992).
21. Milton, Donald, et. al. Endotoxin Measurement: Aerosol Sampling and Application of a New Limulus Method. *American Industrial Hygiene Association Journal.* 51(6):331-7.
22. Reynolds, Stephen J. and D. Milton. Comparison of Methods for Analysis of Airborne Endotoxin. *Applied Occupational Environmental Hygiene.* 8(9):762.
23. Milton, Donald, et. al. Endotoxin Measurement: Aerosol Sampling and Application of a New Limulus Method. *American Industrial Hygiene Association Journal.* 51(6):333.
24. Burge, Harriet A. *Bioaerosols.* Lewis Publishers, Boca Raton, Florida, 1995. p. 83.
25. Burge, Harriet A. *Bioaerosols.* Lewis Publishers, Boca Raton, Florida, 1995. p. 82.
26. ACGIH. *Bioaerosols: Assessment and Control.* ACGIH, Cincinnati, Ohio, 1999. p. 23-10.

Section III

CHEMICALS

Chapter 8

VOLATILE ORGANIC COMPOUNDS

In 1995, the National Institute for Occupational Safety and Health (NIOSH) reported that 17 percent of all their indoor air quality surveys have identified volatile organics as either the cause or contributor to the indoor air quality complaints. Many offices spaces have residual organic components in the air from construction, renovation, maintenance, janitorial, chemical usage/processing (e.g., spray painting associated with marketing projects), and pest control activities. There is also off-gassing from new furnishings, building materials, and office supplies/equipment. Some of the organics may originate from the growth of microbes. Tobacco smoke, deodorants, and perfumes contribute to the total organic loading.

Some indoor chemical contaminants originate outdoors. Outdoor contaminants enter the indoor air predominantly through the fresh air intake but may also enter through structural penetrations or porous structural surfaces. There may be an activity or activities involving chemicals within the building whereby the chemicals are exhausted on the roof, entrained in the air currents, and re-enter the building through the fresh air intake. This is not as frequently encountered as chemicals emitted from other sources in the vicinity of the building.

Automobile exhausts and industrial pollutants prevail in large cities and around industrial plants. Even food manufacturing operations have been known to generate organic chemicals. Environmental organic compounds also evolve from nature's store of plant life.

Industrial activities generate organic air pollutants both inside and outside. Although most of these chemicals are known, some are by-products of multiple-chemical processing and chemical treatments. Stack exhausts may service several areas with different chemical contributions, and complex chemical reactions in the stacks result in complex chemical mixtures in ambient air.

Incidents of chemical storage fires or petrochemical explosions result in the release of unknown organic by-products. Fires in transport systems and buildings result in the release of unknowns. The possibilities are infinite!

Volatile organic compounds include all organic compounds with up to twelve carbons in their molecular structure. They are organic compounds, containing at least one carbon molecule in their structure, and they evaporate at normal pressures and temperatures.

HEALTH EFFECTS AND OCCURRENCES

Industrial exposures to volatile organic compounds (VOCs) are generally ten to one hundred times that of non-industrial home and office environments. Home and office environments are typically two to one hundred times higher than that found outside. A reasonable line of logic would dictate that industrial exposures would result in more health complaints than home and office exposure. Yet, this is not the case.

Many environmental professionals ascribe the complaints to exposures to a medley of chemicals and to the lack of adequate dilution in indoor air. The chemicals are trapped and recycled with the close environments.

Office environments generally consist of up to three hundred different chemicals, an amalgam that certainly complicates an investigation. These VOCs may originate from one, or a combination of, the following:

- Ambient outside air (e.g., methane is reported in ambient outside air at levels of around 1.8 ppm)
- Off-gassing of chemicals from furnishings (e.g., formaldehyde from desks made of particleboard)
- Emissions from office equipment (e.g., toners from copy machines)
- Cleaning and maintenance products
- Construction, demolition, and building renovation activities (e.g., painting the walls)
- Personal hygiene products (e.g., perfume)
- Pesticides and insecticides
- Environmental air pollution (e.g., automotive exhaust)
- Commercial activities (e.g., automotive painting, roof asphalting, and dry cleaning)
- Industrial exhausts (e.g., particleboard manufacturing)

For a list of some of the chemicals frequently found in indoor air quality, and their sources, see Tables 8.1 and 8.2.

Although the health effects of VOCs are chemical dependent, the effects of low level, non-industrial exposures generally found in indoor air quality are relatively consistent. They may involve one, or a combination of, the following:

- Irritation of the eyes, nose, and throat
- Headache
- Lightheadedness
- Nausea

Symptoms resulting from industrial and commercial exposures involve considerably higher exposures to a wide variety of known chemicals. The health effects are more

Table 8.1 Common Indoor VOCs and Their Sources

Pollutant	Indoor Sources
Formaldehyde	Hardwood plywood, adhesives, particleboard, laminates, paints, plastics, carpeting, upholstered furniture coveings, gypsum board, joint compounds, ceiling tiles and panels, non-latex caulking compounds, acid-cured wood coatings, wood paneling, plastic/melamine paneling, vinyl floor tiles, parquet flooring
Benzene	ETS, solvents, paints, stains, varnishes, fax machines, computer terminals and printers, joint compounds, latex caulk, water-based adhesives, wood paneling, carpets, floor tile adhesives, spot/textile cleaners, Styrofoam, plastics, synthetic fibers
Carbon Tetrachloride	Solvents, refrigerant, aerosols, fire extinguishers, grease solvents
Trichloroethylene	Solvents, dry-cleaned fabrics, upholstered furniture covers, printing inks, paints, lacquers, varnishes, adhesives, fax machines, computer terminals and printers, typewriter correction fluid, paint removers, spot removers
Tetrachloroethylene	Dry-cleaned fabrics, upholstered furniture coverings, spot/textile cleaners, fax machines, computer terminal and printers
Chloroform	Solvents, dyes, pesticides, fax machines, computer terminals and printers, upholstered furniture cushions, chlorinated water
1,2-Dichlorobenzene	Dry cleaning agent, degreaser, insecticides, carpeting
1,3-Dichlorobenzene	Insecticide
1,4-Dichlorobenzene	Deodorant, mold and mildew control, air fresheners/ deodorizers, toilet bowl and waste can deodorizers, mothballs and moth flakes
Ethylbenzene	Styrene-related products, synthetic polymers, solvents, fax machines, computer terminals and printers, polyurethane, furniture polish, joint compounds, latex and non-latex parquet flooring
Toluene	Solvent, perfumes, detergents, dyes, water-based adhesives, edge-sealing, molding tape, wallpaper, joint compounds, calcium silica sheet, vinyl-coated wallpaper, caulking compounds, paint, carpeting, pressed-wood furnishings, vinyl floor tiles, paints (latex and solvent-based), carpet adhesives, grease solvents
Xylene	Solvents, dyes, insecticides, polyester fibers, adhesives, joint compound, wallpaper, caulking compounds, varnish, resin and enamel varnish, carpeting, wet-process photocopying, pressed-wood products, gypsum board, water-based adhesives, grease solvents, paints, carpet adhesives, vinyl floor tiles, polyurethane coatings

Table 8.2 Listing of Potential Compounds for Evaluation in Indoor Air Quality Investigations

Compound	Limits of Concern (mg/m^3) ACGIH	WHO[A]	NIOSH	Suspect Source(s)	Relative Response FID[B]	PID[C]
Pentane	1770	—	354	natural gas	65%	8.40@10.6eV
Hexane	176	—	176	rubber cement	75%	4.30@10.6eV
Cyclohexane	1030	—	1930	solvent	85%	1.40@106eV
Decane	—	—	—	copy toner	75%	1.40@10.6eV
Benzene	32	—	0.015C	paints/gasoline	150%	5.30@10.6eV
Toluene	188	375	375	paints/gasoline	110%	0.50@10.6eV
Xylene	434	435	435	paints/gasoline	115%	0.43-0.59@ 10.6eV
Limonene	—	560	—	lemon-odor cleaner	—	—
Acetone	1780	—	590	solvent	60%	1.10@10.6eV
2-Butanone (MEK)	590	—	590	paints/solvent	80%	0.86@10.6eV
Methyl isobutyl ketone	205	—	205	resins/solvent	100%	0.80@10.6eV
Tetrahydrofuran	590	—	590	plastic pipe cleaner	—	1.70@10.6eV
Methyl cellosolve (2-methoxyethanol)	16	—	0.3	solvent/cleansers	—	2.40@10.6eV
Butyl cellosolve (2-butoxyethanol)	121	—	24	solvent/cleansers	—	1.20@10.6eV
Cellosolve (2-ethoxyethanol)	18	—	1.8	solvent/cleansers	—	2.40@10.6eV
Carbon tetrachloride	31	—	12 (1h)	solvent/cleansers	10%	1.70@11.7eV
Tetrachloroethylene (perchloroethylene)	170	—	3	solvent/cleaners	70%	0.57@10.6eV
1,1,1-Trichloroethane	1910	—	1910C	office partitions	105%	0.98@11.7eV
Freon 113	5620	—	5629	coolant	90%	—
n-Nonane	1050	1050	—	not stated	90%	1.40@10.6eV
Methylene chloride	174	350	lowest feasible	not stated	90%	0.89@11.7eV
Trichloromethane (chloroform)	49	270	10	not stated	65%	3.50@11.7eV
1,4-Dichlorobenzene	60	450	10	not stated	113%	0.47@10.6eV

List excerpted from *Evaluation of Sampling and Analysis Methodology for the Determination of Selected Volatile Organic Compounds in Indoor Air.*[1]
Italicized items excerpted from the World Health Organization (WHO) list in ASHRAE Standard.[2]
A Group consensus of concern level as posed by the WHO.
B As calibrated to methane.
C As calibrated to isobutylene.

pronounced and chemical specific. If these higher exposure levels are likely (e.g., exhausted chemicals from a manufacturing operation), the health effects may be more extensive. The investigator should assess impact based on the known health effects of the suspect chemical(s) identified during the preliminary assessment.

People who are chemically sensitive, elderly, infants, and chronically ill will also require special consideration, especially if exposures are 24-hour (e.g., residences and nursing homes). These people are not normally located in office and other work environments.

SAMPLING STRATEGY

A clear, concise air sampling strategy is as important or more so than the actual sampling. If samples do not represent the complaint times, area, and conditions, there is little point in taking a sample. It may appear obvious to many that the when, where, and how are only logical. Yet, logic is sometimes illusive.

When to Sample

At a minimum, samples should be taken during a time or times when complaints are their greatest and for a period of time sufficient to capture adequate sample. This may sound like a simple concept to many readers, yet investigators continue to take samples only during periods deemed most convenient. Sometimes these convenient times do not fall within the time period when people are complaining.

Where to Sample

Identify an area or areas central to where complaints have occurred. After determining the area of concern, determine if there is an indoor, non-complaint area. Perform sampling in an indoor control area and outside at the fresh air intake. An indoor control area may be a non-complaint area within an office building, a manufacturing area associated with office spaces where office occupants are complaining that symptoms do not occur in the manufacturing area, or any of a number of other potential scenarios whereby a comparison may provide useful information. Although they may not be readily apparent and are certainly next to impossible to identify without questionnaires, a concerted effort should be made to identify a control area in all cases. For some examples of where control locations have provided usable background information, see Figures 8.1 through 8.4.[3]

Unless the circumstances indicate otherwise, samples should be taken of areas, not people who are wondering in and out of a complaint area. This is area sampling, not OSHA personnel sampling.

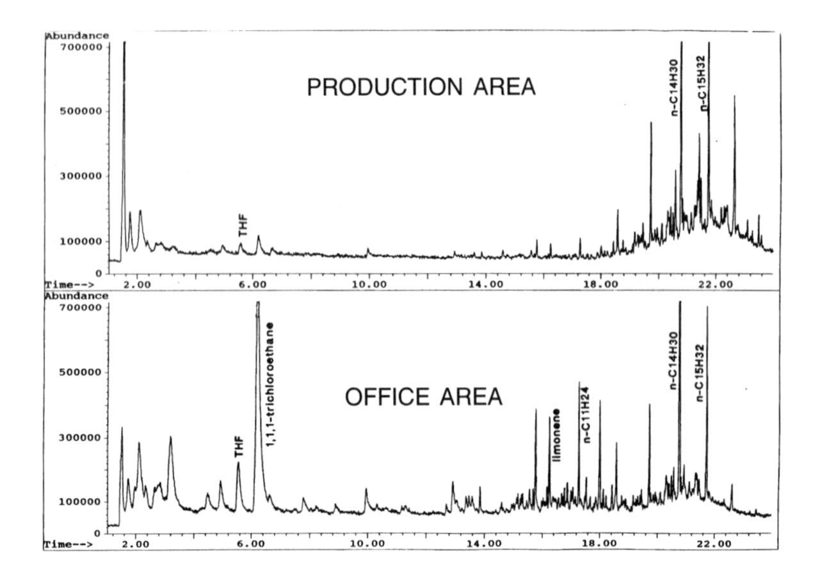

Figure 8.1 Point Source Processing vs. Adjacent Office Area. Organics in office space mimic production area. (Courtesy of NIOSH, Cincinnati, OH)

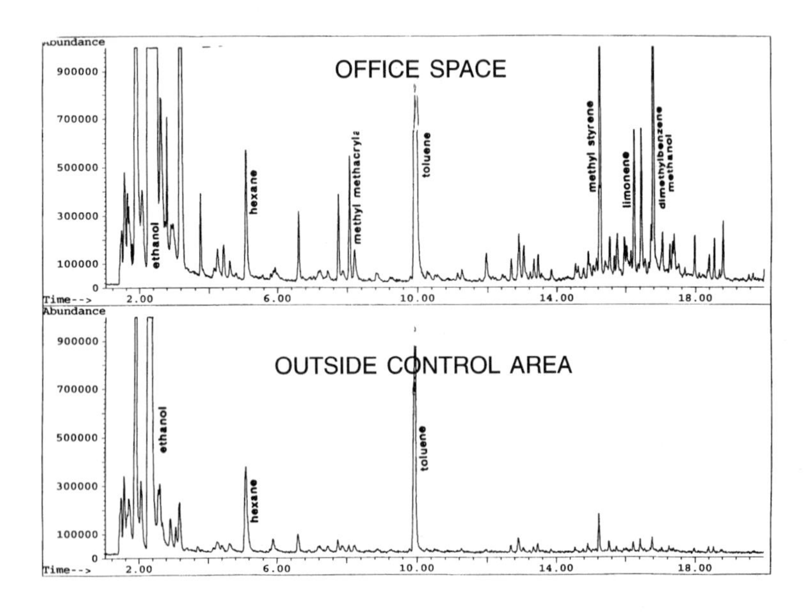

Figure 8.2 Office Space vs. Outside Control Area. Control shows many of same organics found indoors. (Courtesy of NIOSH, Cincinnati, OH)

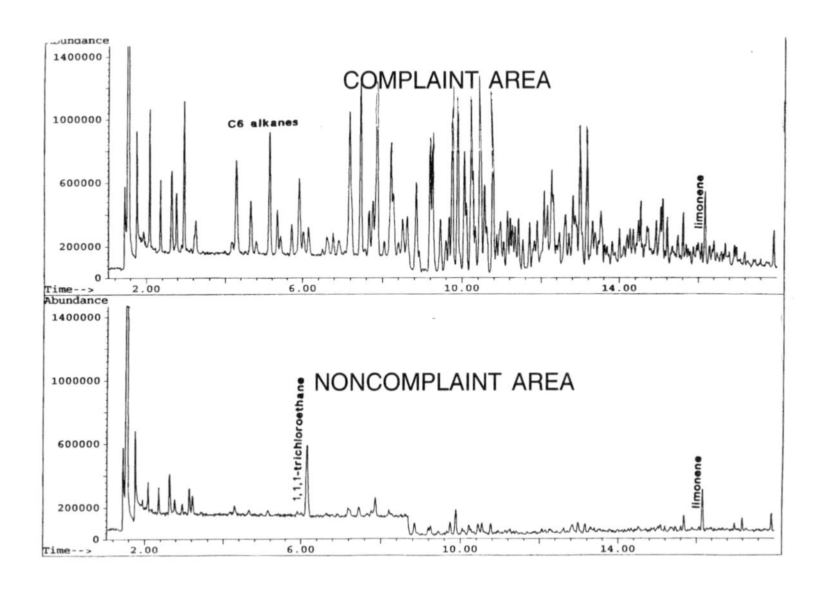

Figure 8.3 Complaint Area at Closet Drain in Office Building vs. Non-complaint Control Area. Speculation was that the source was petroleum distillates, not gasoline from cars. (Courtesy of NIOSH, Cincinnati, OH)

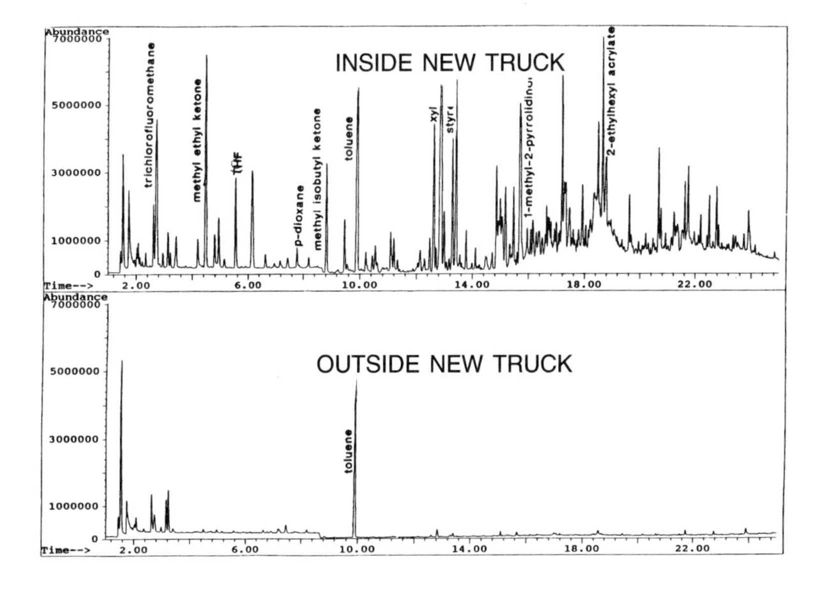

Figure 8.4 At the Source vs. Remote to the Source. The origin of toluene component was outside the truck. (Courtesy of NIOSH, Cincinnati, OH)

Placement of the sampler should be within the breathing zone of those occupying the area of concern, not in a corner at ceiling height. The latter has been observed!

Observations and conditions (e.g., proximity of air supply units and potential source emitters, temperature, humidity, and other observations) should be noted. If blowing directly on the sampler, the effects of the air supply should be recorded. Stagnant air pockets may impact sample results, and proximity of equipment (e.g., copy machines) and activities (e.g., glue application) may prove to be important information upon final review.

Where there is a suspect source, bulk sampling may also be performed of products in question and compared with the air samples. In this case, the data may assist in source identification. For an example, see Figure 8.5.[2]

How to Sample

There is no one-size-fits-all technique for sampling volatile organics. As a single panacea does not exist, the investigator needs to become familiar with the various methods of sampling.

Identification of unknowns involves a lot more expertise and expense. Yet, there may be a surprise that could otherwise not have been anticipated. For instance, a large-scale laundry facility was recycling chlorine-containing water with the detergent and chemicals in the steam treatment process. The end product was chloroform.

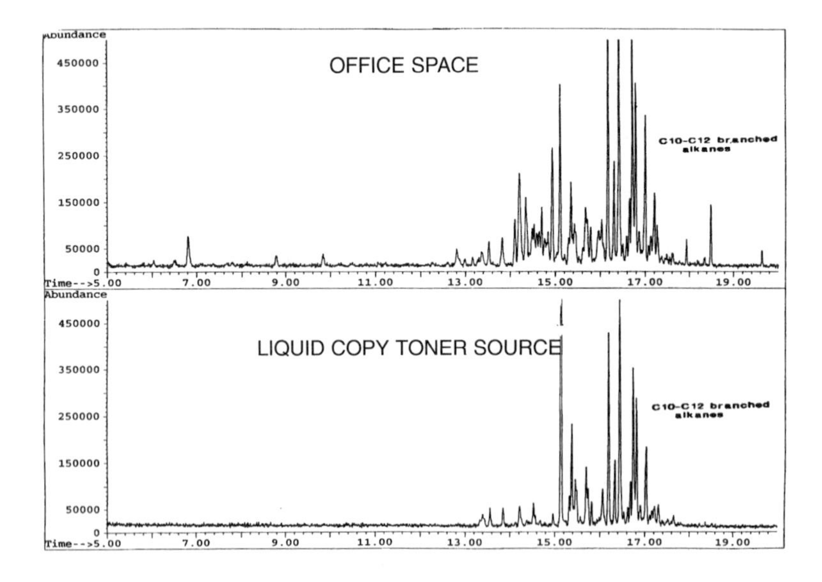

Figure 8.5 Office Space vs. Bulk Liquid Copy Toner. Most of the organic components were due to components in liquid copy toner. (Courtesy of NIOSH, Cincinnati, OH)

Table 8.3 Compounds Used to Challenge Subjects in the Denmark Study[4]

n-Hexane	n-Pentanal
n-Nonane	n-Hexanal
n-Decane	Iso-propanol
n-Undecane	n-Butanol
1-Octane	2-Butanone
1-Decene	3-Methyl-3-butanone
Cyclohexane	4-Methyl-2-pentanone
3-Xylene	n-Butylacetate
Ethylbenzene	Ethoxyethylacetate
1,2,4-Trimethylbenzene	1,2-Dichloroethane
n-Propylbenzene	
a-Pinene	

SCREENING CONSIDERATIONS

Screening procedures for volatile organic compounds are used for determining need for more extensive, costly approaches. The cost for the quantification of total organics is about one-tenth that of identification, and if a worse-case complaint area is assessed, the number of samples to be analyzed can be minimized, again keeping down the overall project cost. If the results are low, additional analytical fees may be circumvented. Yet, the all-consuming question involves the definitions of low and acceptable risk.

There are no established acceptable limits for total organics. Thus, the environmental professional must decide on an action limit to serve as a go-no-go prior to proceeding with the expense of identification and more extensive sampling.

Some environmental professionals choose to use an action level of the lowest ACGIH limit for specific VOCs. The lowest ACGIH limit for volatile organics is that of benzene (0.5 ppm, or 1.6 mg/m^3). Yet, irritation is the main complaint in indoor air quality situations, and irritation levels are sometimes lower than the exposure limits.

One researcher recommends a limit as low as 0.25 mg/m^3, based on irritation response levels and safety factors.[5] Sixty-two chemically sensitive subjects were challenged with twenty-two compounds that were thought to represent compounds frequently found in indoors The researcher identified irritation levels to organic compounds at 5 mg/m^3 (compared to toluene) and gave this number a 50 percent safety factor. The chemicals to which the chemically sensitive subjects were exposed are listed on Table 8.3.

Another researcher recommends a limit of 0.30 mg/m^3 with a limit of no more than 0.06 mg/m^3 for each component.[4] ASHRAE recommended setting a limit of one-tenth the ACGIH limit in its 1989 publication. If the latter were used, the lowest limit would be less than 0.05 ppm, or 0.16 mg/m^3.

Some state agencies set their own in-house limits (e.g., Texas General Services Commission limit: 0.5 mg/m^3). The state of Washington "East Campus Plus Indoor

Table 8.4 Characterization of Common Volatile Organics

Aromatic Hydrocarbons	Chlorinated Hydrocarbons
benzene	methylene chloride
toluene	1,1,1-trichloroethane
o-, m-, p-xylene	perchloroethylene (tetra-chloroethane)
Aliphatic Hydrocarbons	o-, p-dichlorobenzenes
n-pentane	1,1,2-trichloro-1,2,2-trifloro-ethane (Freon)
n-hexane	Terpenes
n-heptane	d-limonene[1]
n-octane	turpentine (pinenes)
n-decane	Aldehydes
Ketones	hexanal
acetone	benzaldehyde
butanone (MEK)	noanal
methyl isobutyl ketone	Acetates
cyclohexanone	ethyl acetate
Alcohols	butyl acetate
methanol	amyl acetate
ethanol	Other
isopropanol	Octamethylcyclotetrasioxane
butanol	
Glycol Ethers	
butyl cellosolve	
diethylene glycol ethyl ether	
Phenolics	
Phenol	
cresol	
2-, 3-, 4-methylphenol	

Air Quality Program" has set a limit for VOCs of 0.5 mg/m^3 for new state buildings. For the general public and commercial establishments, most state limits are guidelines only.

SAMPLING AND ANALYTICAL METHODOLOGIES

Air sampling for unknowns is most effectively performed through screening methods. Screening is mostly used where there are health complaints, and it is unclear that organic chemicals are the cause.

Screening should, however, be avoided when sampling for unknowns, the cost is of little or no concern, and turnaround time is important. For example, a chemical warehouse fire may result in the release of toxins into a residential area. Sampling should be performed at the time of the fire or as soon thereafter as possible and components identified immediately. Results may impact evacuation plans and litigation.

If, however, the organic compounds are known or specific compounds are suspect, screening is unnecessary. Sampling may be performed for the specific chemical(s) for which there is an expressed concern.

Both the EPA and NIOSH have published air sampling methods for known and unknown chemicals. NIOSH has published a method for screening as well. Research is ongoing, and some laboratories develop their own protocols. Only the two government publications that have gone through extensive review processes are presented herein.

The investigator should become familiar for each of the air sampling methodologies. Each method has its own special use and limitation. Its use should be case dependent.

Screening Protocols

The screening procedures target the level of airborne total organics, or hydrocarbons. Sampling may be performed by solid sorbent sampling, evacuated container sampling, or direct reading measurements. The solid sorbent method is a published method, and analysis is by gas chromatography with a flame ionization detector.

Solid Sorbents

The solid sorbent sampling includes the capture of organic components using activated charcoal, extraction (or desorption), and analysis. The following is in accordance with NIOSH 1500, Method for Hydrocarbons:

- Air sampling pump
- Capture medium: coconut shell charcoal
- Flow rate: 0.2 liters per minute
- Minimum air volume: 2 liters
- Maximum air volume: 2-30 liters
- Two to ten field blanks per set (typically one field blank is taken and a laboratory blank is used)
- Desorption: carbon disulfide
- Standard: analyte sought (most laboratories use hexane for total hydrocarbons)
- Analysis: GC-FID
- Limit of detection: 0.001 to 0.01 mg/sample

The laboratory reference standard generally used to determine quantity of total organics is hexane. Laboratory preferences vary. Some may use methane, and others may use a mixture of organics. True quantitation can only be obtained by comparing each component compound with its own. Thus, as the standard is rarely the same as the chemicals in the sample, quantitation of total organics only provides a "ballpark number."

Evacuated Air Containers

Evacuated containers are generally easy to use and have a long shelf life (e.g., can be store for up to a year). Whereas the sorbent samples are limited in capture, the evacuated containers collect all organics, irrespective of chemical characteristics. They include a canister (e.g., SUMMA® canister), bag (e.g., integrated bag sampler), can (e.g., MSA Evacuated Can), and test tube (e.g., Texas Research Institute IAQ Sampler). See Figures 8.6, 8.7, and 8.8.

Evacuated canisters are thermally treated containers under a vacuum. Air samples are collected by opening a valve that is later closed after a pre-designated time period, as little as 1 minute. The canister is metal and will not collapse during shipping, and one canister can be used for screening and sampling for unknowns simultaneously. Although total hydrocarbons can be processed from canisters, very few laboratories are willing or capable of just running a simple scan for total hydrocarbons. As multiple analyses can be performed on one, evacuated canisseconds), record the pertinent information (e.g., sample number and location), secure the opening, and ship to the laboratory. Evacuated cans can be used only for screening.

Ambient air test tubes are "test tubes" with a Teflon® cover and a screw-on cap. To use the ambient air test tubes, unscrew the cap and leave it open in the sample area for 15 to 30 minutes. Waving the tube through the air can reduce required exposure time. Cap the test tube, and send it to the laboratory for analysis.

Analysis by Gas Chromatography

Gas chromatography (GC) is used for quantitation of the desorbed chemicals or of the bulk air samples. Where a sorbent sample is being analyzed, an aliquot of the desorbing solvent and desorbed material is injected into the instrument. Where bulk air is analyzed, the sample is taken by syringe and injected directly into the GC.

An inert gas carries the sample through a column that separates the components. This separation is based on the boiling point of the individual components and on their affinity for the packed material or coating on the column. Then, as they emerge from the column at different rates, the separated components are passed through a detector. In most cases, the flame ionization detector is used.

The flame ionization detector (FID) that detects most organics is the most commonly used detector for analysis of total hydrocarbons. Its response is greatest to simple hydrocarbons (e.g., propane), decreasing with increased substitution of other

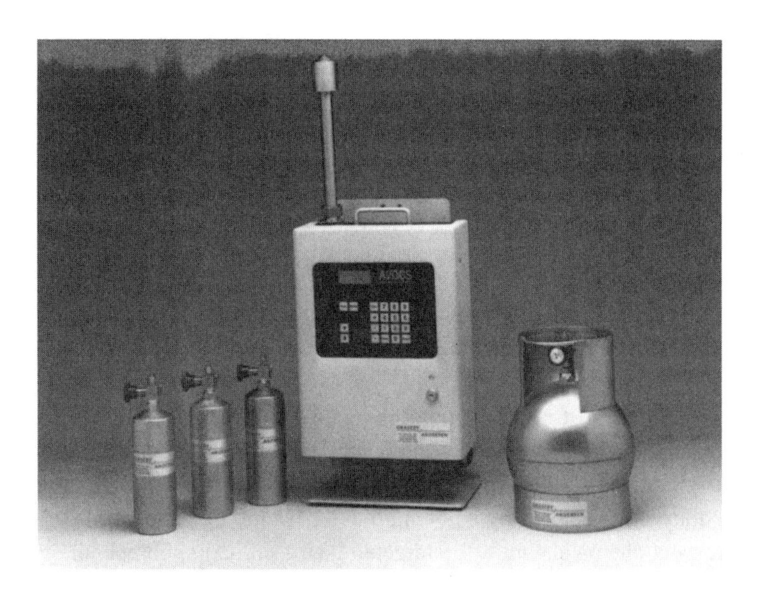

Figure 8.6 SUMMA® Canister
(Courtesy of Graseby Andersen, Atlanta, GA)

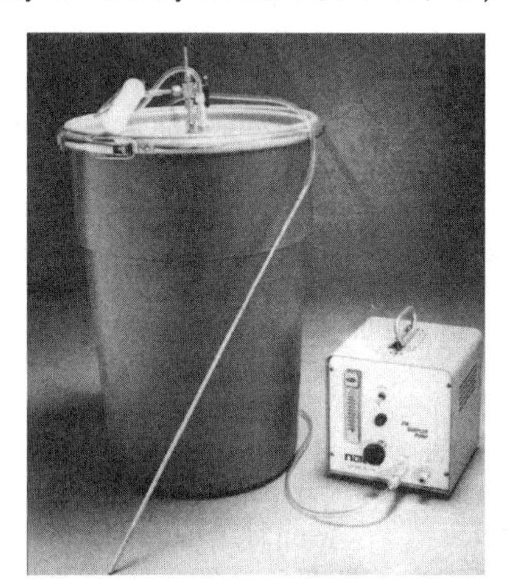

Figure 8.7 100-Liter Integrated Bag Sampler
(Courtesy of Graseby-Nutech, Durham, NC)

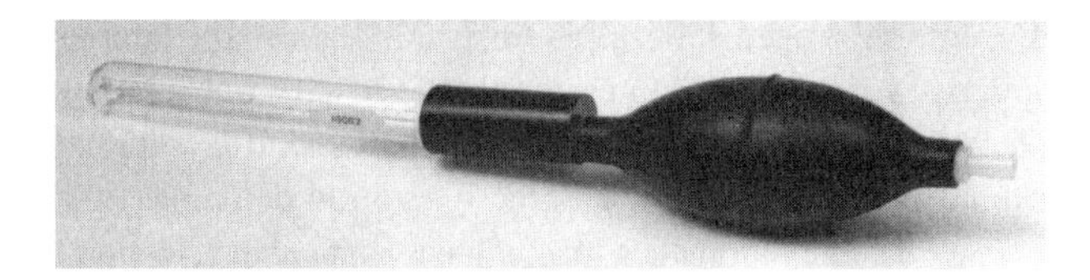

Figure 8.8 Ambient Indoor Air Sampler
(Courtesy of Texas Research Institute, Austin, TX)

elements (e.g., oxygen, sulfur, and chlorine). As the complexity increases, there is a decrease in detection sensitivity. For this reason, most organics will be detected, but quantitation response is diminished with increased complexity.

For GC-FID, the standard is generally methane or hexane. This varies by laboratory. Upon direct air injection, detection by FID is 1 mg/m^3, as compared to methane, or 6 mg/m^3, as compared to hexane. Greater detection below 0.5 mg/m^3 is obtained only through solid sorbent sampling.

Sampling for Suspect or Known Organic Compounds

The preferred, most reliable sampling technique, both qualitatively and quantitatively, occurs where the organic compounds have already been identified. Either specific compounds are known to be present (e.g., toluene from recently applied paint), or they are suspect (e.g., potential 1,1,1-trichloroethane from office partitions). Suspect chemicals are often targeted based on odors, complaint symptoms, and/or probable sources.

Air sampling for organic compounds is generally performed by drawing air through a solid sorbent where the chemical is captured. Solid sorbents are any of a number of solid materials which can capture, or adsorb, specific compounds that can later be extracted, or desorbed, from the collection medium. The sorbent captures, and another chemical desorbs the captured organic compound(s). The desorbed material is then analyzed by laboratory instrumentation. The premise is simple, yet the investigator should be aware a few problems are associated with this sampling approach.

There is no one single sorbent which can be used to sample for "all organics." The most commonly misused sorbent is activated charcoal. Yet, activated charcoal only captures moderately volatile, nonpolar organics (e.g., toluene). It does not capture highly volatile, nonpolar organics nor does it capture polar organics. The misuse of activated charcoal to sample for all organics is a very common practice. Sorbent information is crucial. See Table 8.5.

Even if captured, an organic compound may not be extracted (i.e., retrieved) from the sorbent. Some desorbing chemicals completely drive the compound(s) from the sorbent, but they will not interfere with analysis. Others can be desorbed only partially. Some organic compounds may require special desorbing chemicals, and special processing generally interferes with the analysis of the other organic compounds. In brief, retrieval of all captured compounds, if all are captured, is rarely feasible.

The most commonly used laboratory instrumentation for analyzing volatile organic compounds is the gas chromatograph (GC) with one of three detectors. Each detector is used for detection and quantitation of specific organic compounds. The most commonly used detector, the flame ionization detector (FID), is used to detect aliphatics (e.g., hexane), aromatics (e.g., toluene), and some alcohols (e.g., isopropanol). It is important that the investigator have a good idea or know specifically the type of organic compound for which air monitoring is to be performed prior to sampling.

Table 8.5 Characteristics of Solid Sorbent Media for Sampling Organic Compounds

Type	Chemical Preferences	Affected by High Humidity
Charcoal	Nonpolar volatile and semivolatile organics	yes
Silica gel	Polar volatile and semivolatile organics	yes
Carbon molecular sieve	Nonpolar highly volatile organics	yes
Tenax porous polymer	Nonpolar semivolatile organics	slight
Porous polymer	Selective for specific components (dependent on type of polymer)	no
Alumina gel	Polar high molecular weight organics	yes
Florisil	Polychlorinated biphenyls and some pesticides	no

Volatile: boiling point between 50° and 150°C.
Highly volatile: boiling point between 50° and 100°C.
Semivolatile: boiling point between 2400° and 400°C.

Solid Sorbents and Their Characteristics

Sorbent "collection efficiency" is affected by temperature and humidity. Increased temperatures result in decreased adsorption, irrespective of the type of sorbent used. High humidity (greater than 80 percent) results in decreased adsorption by certain sorbents for some analytes.[6] Those sorbents that are most likely to be impacted by high humidity are charcoal, silica gel, carbon molecular sieve, and alumina gel.

Sorbent capture efficiency is based on known sampling parameters as well. These include the following:[7]

- *Flow rate*—Increased flow rates result in decreased adsorption efficiencies. A moderate flow rate for most organics is at or less than 200 milliliters per minute.
- *Concentration*—Elevated concentrations of competing organics will increase the sorbent loading, resulting in breakthrough. Breakthrough is the passage of a chemical or chemicals through the sorbent without being captured or after capture being displaced by another chemical for which the sorbent has a greater affinity.
- *Sample volume*—While increasing the analytical efficiency, high air volumes will also increase the chances of sorbent loading. The amount of air sampled should be based on the concentration anticipated, the limits of the sorbent, and the limits of the analytical methodology.
- *Competition between chemicals*—Chemicals have a differing affinity in the different adsorbents. Those with a stronger affinity will displace those with a lesser attraction.

- *Sorbent particle size*—The smaller the adsorbing particles the greater the surface area, therefore there is an increase in adsorption.
- *Amount of sorbent*—An increase in sorbent volume will result in an increase in capture material. Sorbent tubes contain between 100 and 1000 milligrams of material.
- *Type of sorbent*—The solid sorbents vary not only in their ability to capture specific compounds but in their feasibility for extraction within the laboratory. Many toxic organic compounds have been tested with the various sorbents for capture and extraction efficiency, but many have not. Particularly difficult are those compounds that are not listed as toxic or that are irritants only. These are likely to have been excluded from any research or development studies.

Sampling Flow Rates and Air Volumes

The air flow of the sampling pump should be calibrated according to published guidelines. Air sampling rates are, however, rarely a single number.

The environmental professional is given a range from which to choose or a maximum flow rate. The high range should not be exceeded under any circumstance due to sample loss. The higher the sampling rate, the greater the chances for breakthrough. Breakthrough results when 10 to 25 percent of the captured chemical in the first section is exceeded in the second section of the sorbent.

The lower end of the range is less important, but it does require a longer sampling time to collect sufficient sample to detect at the low levels needed for indoor air quality investigation. Yet, extremely low sampling rates may result in diffusion uptake rates that are at or below 5 cubic centimeters per minute.

All choices in between are discretionary. Relevant issues, which may be used to adjust these choices, include the following:

Desired sample duration—If exposures are anticipated within an abbreviated time period, that window of occurrence will provide information regarding a worse case scenario. Where exposures are thought to be consistent throughout the day, sampling may be performed within that time period. Where exposures are thought to be consistent but a quick sample is desired, the duration may be reduced accordingly.

Desired air volume—Higher anticipated sample volumes will require adapting the flow rate, along with the sample duration, to assure a sufficient amount of air has been sampled. For short duration samples, requiring large volumes of sampled air, the air flow rate will need to be at the top end of the range. For longer durations, requiring small air volumes, the low end of the range is indicated.

Desired air volume is based upon a combination of the sampling methodology limits and on the anticipated concentration of organics in the air. The sampling methodology limits are published or provided by the laboratory. The anticipated concentration of organics can, however, be a crapshoot. Unless there are visible emissions, odors that are strongly evident to all passers-by, or the point source is known to be emitting copious amounts of chemical, concentrations are likely to be low. If monitoring an industrial process, stack emissions, or hazardous waste emissions, concentrations are likely to be high. Professional judgment is necessary.

Once the parameters have been decided, set the pump(s), collect the desired air volume, and send the sample(s) to the laboratory of choice, along with the sampling data. Sampling data should include the following:

- Sample name and/or number
- Air volume sampled
- Average temperature
- Relative humidity
- Barometric pressure

Care should be taken so as not to contaminate sampled sorbents during shipping. Contamination may occur from sources in the packing material [e.g., a bulk organic sample shipped with the sorbent tube(s)] or from outside (e.g., a chemical spill by the freight company). Bulk organic samples should never be shipped with sorbent tubes, and all samples should be accompanied by at least one blank. A blank is a separate opened, capped sample that has not been sampled.

Analysis by Gas Chromatography

At the laboratory, the sample is desorbed (i.e., extracted). The extracted chemical is then injected into an analytical instrument. Only about one thousandth of the chemical actually collected is analyzed. The remaining desorbed chemical is then stored for a defined period of time (e.g., typically about one month). Thus, additional analyses can be requested.

Helpful Hints

There are published methodologies for most of the more common organic compounds. They are discussed under the heading of Published Sampling and Analytical Methodologies. Become familiar with the principles contained herein and seek technical support. Many analytical laboratories provide a client service department whose sole responsibility is to assist their clients.

Searching for Unknowns

Identification of unknowns is performed by gas chromatography/mass spectrometry (GC/MS). This procedure combines the quantitative and chemical separation features of the gas chromatograph with chemical unknown identification. The mass spectrometer takes the chemical components that have been separated by the gas chromatograph, fragments each separate unknown, and records the fragmentation pattern. This pattern is, in turn, compared with features of known chemicals. The comparison is typically made by a computerized library search of other patterns. Then, the computer provides information as to the best fit for each of the components. Rarely is the match one hundred percent, while ninety-five percent is considered a very good fit. The chemist then decides if the match is significant for identification or whether the chemical cannot be identified through the library. Most GC/MS libraries have over 70,000 chemicals on file. Some have as many as 150,000. These should identify most of the more common or widely encountered chemicals.

Although the cost for GC/MS analyses is high, taking only one sample, in order to avoid the additional cost for analysis of a control sample, may lead to inconclusive or poorly substantiated results. Without an acceptable standard, a control can serve as a gauge, or means for comparison.

Although the analytical method for identification of unknowns is similar, the collection methods are variable. The investigator may use one or a combination of methods, depending on the circumstances. These methods include multibed sorbents, evacuated samplers, and air sample bags.

Multibed Sorbent(s) with Thermal Desorption

Multibed sorbents concentrate large sample volumes into one tube. Where a chemical is present in extremely low quantities, it may not be detected without concentrating the sample onto a sorbent.

The more versatile the sorbent(s) the better the collection efficiency. The objective here is to collect a broad range of chemicals and be able to desorb as many of the captured chemicals as possible. This is best accomplished by thermal desorption.

With thermal desorption, the sorbent is heated while attached in-line with the analytical equipment. As the temperature is increased, the captured organic compounds are driven off from the sorbent.

In a study performed by NIOSH, the effectiveness of thermal desorption over chemical desorption was clarified. Two separate sorbents were used to sample within the same time period and same sample site (e.g., a rubber molding facility). One was sampled by thermal desorption tube with a carbon based sorbent and analyzed by GC/MS. The other was sampled by charcoal tube, chemically desorbed, and analyzed by GC/MS. The results favored thermal desorption sampling and analytical approach over the other.[2] Compounds missed in the chemically desorbed charcoal tube included aliphatic amines, sulfur dioxide, and carbon disulfide. As it is used as the desorption compound for most chemical extraction from charcoal tubes, carbon disulfide would have been overlooked as an indoor air contaminant had chemical desorption been used. See Figure 8.9.[3]

In another study, NIOSH evaluated the capture and analysis of the component sorbents that they used to composite multibed sorbents. These were, in order of occurrence (e.g., direction of sample flow), Carbopack Y, Carbopack B, and Carboxen. The first layer captured the least volatile compounds. The second captured more volatile compounds, and the third captured those with the highest volatility. Some layers collected portions of that which was collected on an adjacent layer. Had only one sorbent been used, the other compounds would have been lost, not identified. See Figure 8.10.[3]

Evacuated Air Canisters *(e.g., SUMMA® Canisters)*

Evacuated air canisters are specialty samplers. One such sampler is referred to as the SUMMA® canister.

Canisters come in various sizes (ranging from 0.85 to 33 liters in volume). The most common size is 6 liters. Each canister has been treated with chrome-nickel oxide internally to prevent rusting and minimize organic adherence to the surface of the container, and the canister can be used either in its evacuated stage or with a sampling pump. Prior to use, each canister must be cleaned and prepared by a laboratory that has canister analytical capabilities.

The canister cleaning process takes up to 24 hours per canister, so the laboratory will require some lead time. As part of the planning process, the opening may also be fitted with a flow control device or the inlet remains as is. It depends on the sample duration desired by the environmental professional. For a 6-liter canister, the valve provided with the canister may be opened, and the sample collection time will take less than 30 seconds. For longer sample durations, the opening may be fitted with a special control valve or critical orifice with a known, calibrated flow rate. These devices will permit the vacuum within the canister to draw the sample air over a time period of up to 24 hours.

To use an evacuated canister after it has been cleaned and fitted with special flow control (when required), the inlet or control valve is opened. The ambient air is drawn into the canister by vacuum. Start and finish times should be recorded along with pertinent sample information and location. If samples are taken over an extended period of time, the temperature and humidity should be checked routinely and averaged for reporting purposes.

The sample locations may be remote to the site of the canister by using airtight seals, extension tubing, and allowing the ambient air to fill the extension tube prior

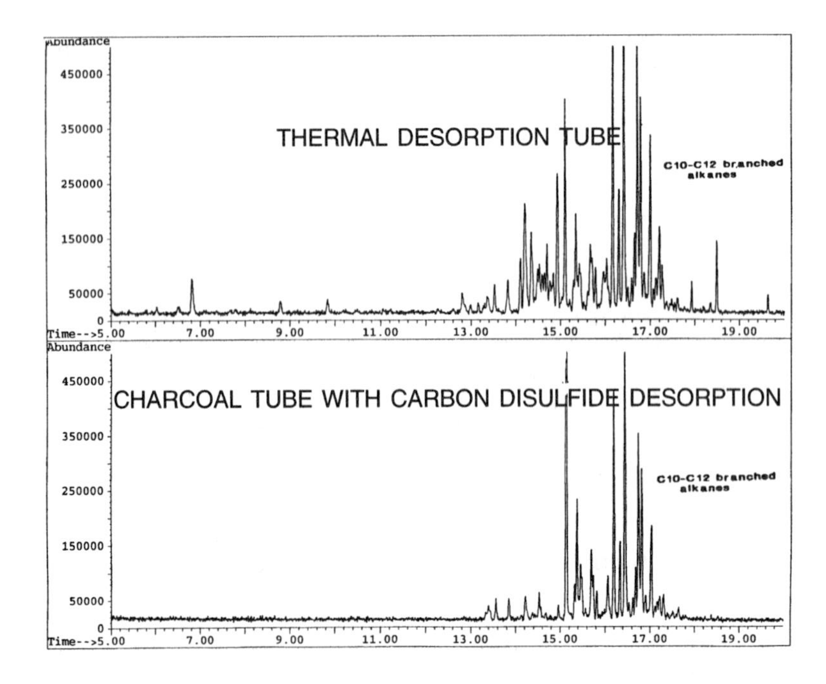

Figure 8.9 Thermal Desorption vs. Chemical Desorption. (Courtesy of NIOSH, Cincinnati, OH)

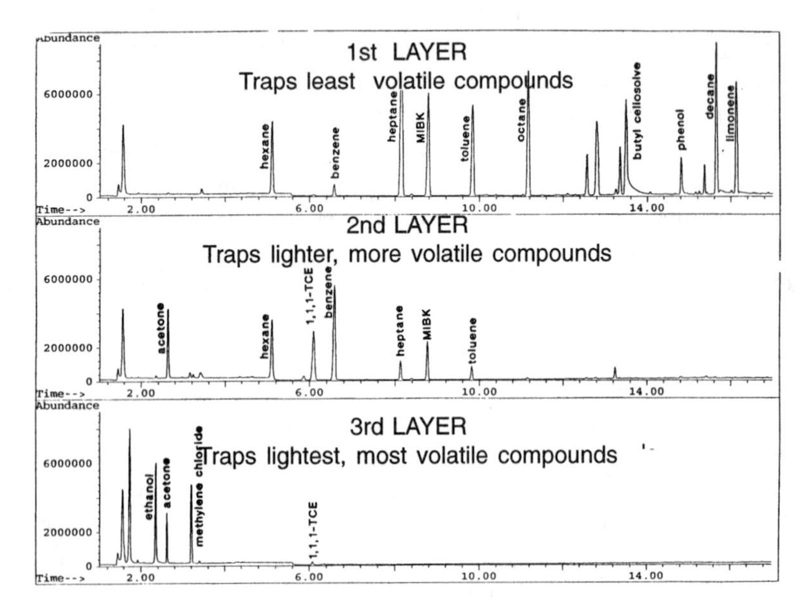

Figure 8.10 Multibed Thermal Desorption. (Courtesy of NIOSH, Cincinnati, OH)

to sample collection. Care should also be taken to choose tubing that is inert and will not off-gas organics. Canister detection is the same as that of ambient air samplers without special cryogenic processing.

Cryogenic processing allows for detection down to 5 ppb (or 0.005 mg/m^3), as compared to methane. This number varies, depending upon the laboratory standard or standards used and the types of components to be identified.

Ambient Air Sampling Bags

Air sample collection bags are special sampling equipment, constructed of any of a number of synthetic materials (e.g., Tedlar® and Teflon®). Care should be taken to confirm that the bag material chosen for sampling will not off-gas materials which the environmental professional may be attempting to capture, components that may interfere with or may be lost through surface adhesion to the sample bag. Commercially available bags range in holding capacity from 0.5 to 120 liters, and sampling does require a vacuum pump.

In order to perform ambient air sampling, a special bag sampling system must be developed or purchased. This involves retrofitting an airtight container (which is larger than the fully-filled bag) with attachments to allow access of the bag valve to the ambient air. Install a sampling pump and tube into the container. Assure all materials used in the construction of the system are not chemically reactive, and test for air tightness. The same system is reusable. There are commercially available integrated bag samplers. See Figure 8.7.

A chemically non-reactive sample bag is installed and sealed within an airtight container which is larger than the fully-inflated sampling bag and has chemically non-reactive, valve-fitted inlet and outlet ports. Where the inlet is connected to the sample bag, the outlet is connected to a vacuum pump. As the vacuum pump draws a vacuum inside the sealed container, the sample bag draws in ambient air to replace the void that is created inside the sealed container. In this manner, the collected air does not pass through a previously contaminated sampling pump, and there is limited contact with various surfaces (to which chemical may adhere) that might result in the loss of chemicals from the air being sampled.

The sample duration may be controlled by the flow rate of the sampling pump, and upon completion the bag valve stem is closed. The container seal is released, and the sample is removed/prepared for shipping. Be sure to record all pertinent information as discussed previously (e.g., sample name or number, temperature, humidity, and barometric pressure).

Shipping poses several problems that are unique to this method. Due to the failed rigidity of the sampling bag, an extreme change in atmospheric pressure (which does occur in air transportation) may result in an expansion of the bag to the point where its integrity is compromised if the bag is full. It is not unusual for sample bags to arrive at their destination with nothing inside. The way to avoid this problem is to collect only half the capacity of the bag, or ship the bag in a rigid, pressurized container.

Detection is the same as that of an ambient air tube (e.g., as low as 0.5 mg/m^3 as compared with a methane standard). Analysis is also performed in a similar fashion.

Analysis by GC-MS

All sampling methods mentioned may be analyzed by direct injection of the air sample into a GC or GC/MS. Not all GC or GC/MS equipment is outfitted to accommodate direct injection of ambient air samples, and rarely is a GC/MS fitted with a special cryogenic concentration of samples.

The amount of actual sample that is injected into the analytical equipment is typically 1 milliliter. In the case of ambient sample tubes and bags, the sample is undiluted. The canister sample may be analyzed in a fashion similar to that of the tubes and bags, but cryogenic processing gives greater detection.

Although the EPA "TO" Method 12 involves the use of an ambient air canister and analysis by cryogenic treatment and injection into the GC/MS, not all commercial laboratories generally provide this service. All bulk air sampling methods may be used for screening of "total organics" by GC or for identification of suspected large quantities of organic components by GC/MS if the laboratory is outfitted to perform the service.

Cryogenic processing of bulk air samples is typically associated with the GC/MS. It is not generally used for screening of "total organics." The cryogenic process takes 100 to 1,000 milliliters of sample and concentrates it prior to sample release for analytical processi ng. As compared with the other direct injection samples, this method secures one hundred to one thousand more sample than the other bulk air samples with "most all of the original ambient air sampled." This is only surpassed by thermal desorption of sample volumes in excess of 1 liter, and the sorbents are highly selective in their ability to deliver a comprehensive sample.

It is worthy of note that even if all the original ambient air is present, humidity impacts the analysis, particularly where a sample is to receive a cryogenic process. Not only are the chemical components concentrated by the process, the water is concentrated as well.

Although water can be filtered from the system, polar compounds have a tendency to be attracted to the water and subsequently get filtered out with the water. They may not be identified. Should the water be allowed to enter the system, it tends to plug the system causing other problems. The lower the humidity, the better the analysis. Where high humidity levels are normal, some areas of the country are poor locations for considering reliable canister sampling.

At the present, there is an EPA list of ambient air reference standards for cryogenically treated canister samples. These are listed in Table 8.6 and have come from EPA Method TO-14.

Table 8.6 List of Chemical Reference Standards in EPA Method TO-14

Chemical Name (in order of retention times)		
Freon 12	Benzene	o-Xylene
Methyl chloride	Carbon tetrachloride	4-Ethyl toluene
Freon 114	1,2-Dichloropropane	1,3,5-Trimethylbenzene
Vinyl chloride	Trichloroethylene	1,2,4-Trimethylbenzene
Methyl bromide	cis-1,3-Dichloropropene	m-Dichlorobenzene
Ethyl chloride	trans-1,3-Dichloropropene	Benzyl chloride
Freon 11	1,1,2-Trichloroethane	p-Dichlorobenzene
Vinylidene chloride	Toluene	o-Dichlorobenzene
Dichloromethane	1,2-Dibromoethane	1,2,4-Trichlorobenzene
Trichlorotrifluoro ethane	Tetrachloroethylene	Hexachlorobutadiene
1,2-Dichloroethane	Chlorobenzene	
cis-1,2-Dichloro ethylene	Ethylbenzene	
Chloroform	m,p-Xylene	
1,2-Dichloroethane	Styrene	
Methyl chloroform	1,1,2,2-Tetrachloroethane	

Helpful Hints

Although thermal desorption of sorbents is typically a one shot affair, depending on the thermal desorption equipment used, further, more extensive analyses can be performed on sorbents, evacuated canisters, and ambient air sampling bags.

Thus, an investigator can use the latter methods for screening as well. A little foresight by the investigator may avoid the need for additional trips to the job site.

Synopsis of Published U.S. Government Methods

The most commonly used methodology for indoor air quality is that which has been published by the Environmental Protection Agency. The National Institute for Occupational Safety and Health has developed a "screening method" for indoor air quality, and they have published chemical specific methodologies that were developed for industrial exposures.

Environmental Protection Agency (EPA)

The EPA has developed the "Toxic Organic Series," referred to as the EPA "TO" Methods. These methods are becoming widely used and are frequently referenced by environmental professionals.

TO Method 14 is associated with specific compounds, and the search is limited to specific compounds which are summarized in Table 8.6. While the search is limited, results are easier to quantitate through comparison with a standard. Then, too, there is an alternative that covers all bases. This alternative is Method 3.

TO Method 3 covers a search for known as well as a search for other significant peaks. Results typically include the listing of chemicals sought under the method, and the level of detection or actual amount of material detected.

In addition to the Toxic Organic Series, the EPA has also published some methods for indoor air quality assessments. These are referred to as the EPA "IP" (Indoor Pollutant) Series. Although the IP-1B Method for Volatile Organic Compounds (VOC) neither restricts nor mandates the search for any specific compounds, the method is limited to the capture and analysis of only nonpolar volatile organics.

Most of the methods require an air sampling pump, collection media, and laboratory analysis. See Table 8.7 for a summary of methods, target analytes, collection media, flow rates, maximum air volumes, detection limits, and analytical methods.

There are several variations of these EPA methodologies that are generated by laboratories and individual environmental professionals. Some laboratories restrict their efforts only to established and published methodologies. Others go the extra mile to accommodate special needs and requests. Laboratories should be checked out prior to developing a sampling strategy.

National Institute for Occupational Safety and Health (NIOSH)

In 1996, NIOSH published an approach for addressing indoor air quality issues. It is referred to as the Screening Method for Volatile Organic Compounds. The approach is as follows:[1]

- Standard: a broad range of volatile organics
- Collection Medium: multiple sorbent thermal desorption tube
- Flow Rate(s): 0.01 to 0.05 liters/minute
- Recommended Sample Volumes: 1 to 6 liters
- Detection Limit(s): 100 nanograms per tube
- Analytical Method(s): thermal desorption by GC/MS

This approach has led to the identification of twenty compounds commonly found in indoor air quality assessments. See Table 8.2 for a listing of the most frequently implicated VOCs in indoor air quality, recommended limits (i.e., ACGIH, WHO, and NIOSH), and exposure sources. Table 8.1 has a more extensive list of chemicals by class that was published under the NIOSH Screening Method 2549.

Since the 1970s, NIOSH has published and continues to update publications regarding identified chemicals (e.g., also performed sorbent and sample placement

Table 8.7 Environmental Protection Agency Air Monitoring Methodologies

Methods	Target Analytes	Collection Material	Flow Rate(s)	Maximum Air Sample Volume	Detection Limit	Analytical Method(s)
TOXIC ORGANICS						
TO-1	40 Nonpolar volatile organics	Tenax thermo-packed	6-500 ml/min.	10-250 liters	0.002 μg report: ppb	TD/GC/MS
TO-2	TO-1 Organics + 25 nonpolar "highly" volatile organics	Carbon molecular sieve thermo-packed	15-400 ml/min.	20-100 liters	0.002 μg report: ppb	TD/GC/MS
TO-3	TO-2 + chemical library search	Carbon molecular sieve thermo-packed	100-200 ml/min.	100 liters	0.002 μg report: ppb	TD/GC/MS
TO-4	Organochlorine pesticides + PCBs	Glass fiber filter +PU F*	200-280 l/min.	40 x 10^4 liters	ng/m^3	GC/ECD
TO-5	Aldehydes & ketones	Dinitrophenylhydrazine (DNPH) solution	100-1,000 ml/min.	80 liters	ppb	HPLC
TO-6	Phosgene	2% Aniline in toluene-	100-1,000 ml/min.	50 liters	< ppb	HPLC
TO-7	Amines	Thermosorb/N tube	100-2,000 ml/min.	300 liters	1 pg/m3	GC/MS
TO-8	Phenol & cresols	Sodium hydroxide solution	100-1,000 ml/min.	80 liters	1 ppb	HPLC
TO-10	Organochlorine pesticides	PUF	1-5 l/min.	5,000 liters	0.01-0.1 μg/m^3	GC/ECD (other detectors)
TO-11	Aldehydes & ketones	DNPH Treated-silica gel sorbent	200 ml/min.	300 liters	1-2 ppb	HPLC
TO-12 organics	Non-methane canister	Ambient air	NA	NA	1 ppb (cryogenic	GC/FID

Table 8.7 Environmental Protection Agency Air Monitoring Methodologies (continued)

Methods	Target Analytes	Collection Material	Flow Rate(s)	Maximum Air Sample Volume	Detection Limit	Analytical Method(s)
TO-13	Polynuclear aromatic hydrocarbons	Quartz fiber filter + PUF (or XAD-2)	200-280 l/min.	325 x 103 liters	0.1-1.0 μg/m^3	GC/FID, GC/MS, Or HPLC
TO-14	41 Volatile organics	Ambient air canister treatment)	NA	NA	0.1 ppb	GC/MS (cryogenic
TO-15	Polar & water soluble volatile organics	—	—	—	—	—
TO-16	Broad spectrum of organics and inorganics	Direct reading field/lab instrument	NA	NA	variable	FTIR
TO-17	Volatile organic compounds	Multibed sorbent	—	—	—	GC/MS
INDOOR POLLUTANTS						
IP-1B	Volatile, non-polar organics	Tenas thermo-packed sorbent	6-500 ml/min.	20-200 liters	0.002 μg reported in ppb	TD/GC/MS
IP-2A	Nicotine	XAD-4 sorbent	1 l/min.	480 liters	0.02 μg/m^3	GC/NSD
IP-6A	Formaldehyde & other aldehydes****	DNPH-treated silica gel sorbent	100 ml/min.	300 liters	1-2 ppb	HPLC
IP-6C	Formaldehyde & other aldehydes	DNPH treated passive monitor	NA	NA	variable	GC/NSD
IP-7	Polynuclear aromatic hydrocarbons	Quartz fiber filter + PUF	20 l/min.	30 x 10^3 liters	<1 μg/m^3	GC/FID, GC/MS, or HPLC

Table 8.7 Environmental Protection Agency Air Monitoring Methodologies (continued)

Methods	Target Analytes	Collection Material	Flow Rate(s)	Maximum Air Sample Volume	Detection Limit	Analytical Method(s)
IP-8	Organochlorine & other pesticides	PUF	1-5 l/min.	5,000 liters	<1 $\mu g/m^3$	GC/ECD (other detectors)

TO Method 17 will eventually replace TO Methods 1 and 2.

*	A PUF is a 3-inch by 6-cm diameter polyurethane foam plug.
**	Iced impingers that contain a two-phase mixture of aqueous DNPH and iso-octane.
***	Impinger with solution of 2% aniline in toluene.
****	May also use Waters Sep-Pak Silica Gel Cartridge at a flow rate of 2 l/min., air volume of 1,000 liters, detection limit of <1 ppb.
NA	Not applicable.
—	Information unknown or unpublished.

Analytical Procedures

GG	Gas chromatography	
	TD/GC/MS	Thermo Desorption/Gas Chromatography/Mass Spectrometry
	GC/MS	Gas Chromatography/Mass Spectrometry
	GC/FID	Gas Chromatography/Flame Ionization Detector
	GC/ECD	Gas Chromatography/Electron Capture Detector
	GC/NSD	Gas Chromatorgraphy/Nitrogen Selective Detector
HPLC	High Pressure Liquid Chromatography	
FTIR	Fourier Transform Infrared Spectrometry	

studies). These are discussed in greater depth within the subsection on Solid Sorbent Sampling.

INTERPRETATION OF RESULTS

Where screening is performed, the investigator decides on an acceptable limit, based on professional judgement. Professional judgement is based on experience and guidelines set forth by others. To recap the guidelines discussed in Screening Considerations, many investigators choose one of the following:

- A limit of 0.5 mg/m^3, as compared to hexane, guideline used by some state regulators.
- A limit of 5 mg/m^3, as compared to methane, guideline used by some consultants where the laboratory uses methane as the standard. This guideline would be slightly less than 1 mg/m^3, as compared to hexane.

The investigator may choose a stricter limit (e.g., 0.16 mg/m^3, or $1/10^{th}$ of the ACGIH limits for any specific chemical) or a less restrictive limit. Once again, the screening procedures are not intended to regulate. Screening is performed to establish a limit whereby further sampling should be performed.

Where the airborne levels of total organics are less than the limit decided on by the investigator, no further sampling or analyses are necessary. If, however, the screening levels exceed the limits, sampling for unknowns should be performed. Once identified, the VOCs may be assessed on the basis of the identified, known chemical(s) and limits set for them.

The acceptable limit for a known chemical in indoor air environments must, once again, be based on professional judgement. The investigator may choose to use any one of the following:

- The ACGIH guidelines for specific chemicals (not recommended for use where 24-hour exposures may occur)
- The NIOSH guidelines for specific chemicals (generally more stringent than the ACGIH guidelines and not recommended for use where 24-hour exposures may occur)
- One tenth of the ACGIH guidelines for specific chemicals
- The WHO guidelines for specific chemicals (see Table 8.1)
- Known sensory irritation levels for specific chemicals
- EPA limits on a few chemicals that have been so assigned

In some cases, investigators have used OSHA sampling techniques and regulatory limits for assessing indoor air quality in offices and residences. Keep in mind, OSHA exposures are intended for industrial work exposures where all chemicals to which the worker is exposed are known, and they are the least restrictive of all the limits. An attempt to use OSHA limits in complex office and residential environments is "not recommended."

Where limits fail, there has been little or no research performed, or there is some degree of uncertainty, comparative samples become important. These include the following:

- Compare problem and non-problem samples
- Compare indoor and outside samples
- Compare previously taken samples in the building when there were no complaints to those taken in response to complaints

In brief, there are no regulatory limits for indoor air quality. Assessing VOCs is complex, and interpreting the results is even more difficult. If inexperienced in assessing VOCs, seek assistance from someone with experience.

DIAGNOSTIC SAMPLING METHODOLOGIES

The VOC saga does not end on a simple note. There are a few more tools that may be added to the investigator's bag-of-tricks. Bulk liquid sampling is used as a used as a backup for tracking and diagnosing sources, and direct reading instrumentation can be used for tracking as well. The investigator may even find other uses not stated herein.

Bulk Liquid Sampling

Bulk samples may be taken for comparison with the air samples. Sometimes they are gases in liquid. Sometimes they are liquid. Other times they are in powder form.

This process is fairly simple. Identify the sample material, and determine if the sample is suspended in water. If in its concentrated, pure form and not suspended in water, a small sample may be taken by using a pre-cleaned, glass container of any size in excess of 5 milliliters. Be sure to fill the bottle to overflowing. Carefully install the cap, or slide the separate Teflon into position, cap, and seal the top. This process minimizes air pockets where organic gases may collect. Two samples of each bulk liquid should be taken—one for analysis, the other as a backup.

If diluted or suspended in water, the sample size may have to be coordinated with the analytical laboratory. Generally, a 1-liter sample will be requested where the sample is suspended in water. Fill and seal by the same technique used for smaller samples.

Analyses of the concentrated sample that does not have water can be performed by GC or GC/MS. As part of the air sample "screening process," a bulk liquid sample may be compared against the air samples for retention time on the GC prior to proceeding to the expense of a GC/MS. This may provide a means to determine if the associated air sample contaminates originated from the unknown liquid. Following the screening process, the contents of the liquid may be identified by GC/MS instead of, or in tandem with, the air sample. In this fashion, identification and source information may thus be obtained simultaneously.

As for those samples that have been diluted or suspended in water, these materials will likely require extraction prior to analysis. The laboratory will want the diluent identified if possible. Be prepared to provide additional information. After extraction, the samples are processed in a fashion similar to the concentrate.

Direct Reading Instrumentation

Once VOCs have been determined as a probable contribution to indoor air quality complaints, a direct reading instrument may be useful in tracking the source(s). In some instances, this may only be done with a direct reading instrument with data logging capabilities. See Figure 8.11 for a graphic example of one such instrument.

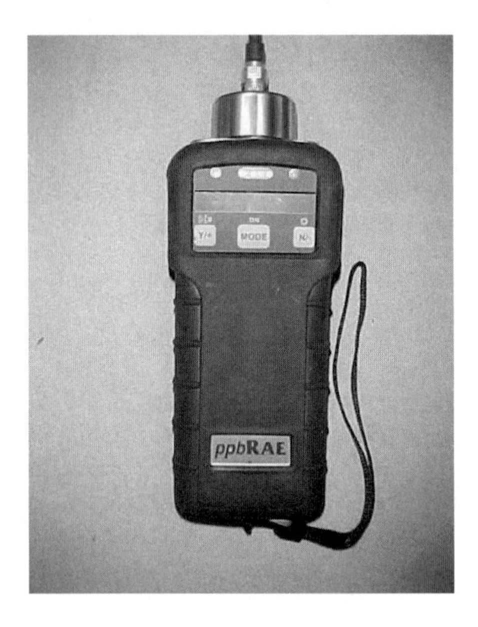

Figure 8.11 Photo of ppbRAE. (Courtesy of RAE Systems, Sunnyvale, CA)

For example, employees working in a gasoline station kiosk were complaining of gasoline odors. A data logging PID (e.g., MultiRAE) was used to determine that exposures were minimal when the windows and/or door were open. When the instrument was left after work when the windows and doors were closed, the instrument recorded a gradual increase of VOCs throughout the evening. When employees opened the door in the morning, the levels dropped back to barely detectable. The source was tracked with the instrument to floor penetrations serving as a conduit for gasoline vapors originating from an apparent leaking underground storage tank. This situation could not have been resolved without the aid of a data logging PID.

There are a couple types of direct reading field instruments. They have different capabilities and limitations, and not all instruments have data logging capabilities. As the choice of instruments may be situation dependent, the investigator should become familiar with each type, know their data logging abilities, and ease of handling. There is nothing more frustrating than having an expensive piece of equipment that is difficult to use.

Flame Ionization Detector

A direct reading instrument with a flame ionization detector (FID) is also referred to as an organic vapor analyzer (e.g., Foxborro® OVA). It responds to low molecular weight aliphatics, aromatics, and some hydrogen-substituted hydrocarbons. The FID has decreasing sensitivity with increased substitution of oxygen (e.g., isopropanol), sulfur (e.g., carbon disulfide), and chlorine (e.g., 1,1,1-trichloroethane). The instrument responds to methane and is not sensitive to effects of humidity.

Photoionization Detector

A photoionization detector is commonly referred to as a PID. A PID responds to most molecular weight aliphatics (except methane), aromatics, many of the hydrogen-substituted hydrocarbons, and some inorganics (e.g., hydrogen sulfide). Although the instrument is sensitive to effects of humidity, a moisture trap can be used when sampling in high humidity environments. Most PIDs have data logging capabilities. Although most PID readings are in ppm, a recently-developed PID reads in the ppb range (e.g., RAE Systems ppbRAE).

The recently marketed ppbRAE is finding a niche in indoor air quality investigations. It has been used in the screening process to track worse-case sample sites for VOC sampling where the levels are less than 1 ppm. The ppbRAE is being test for mold VOC tracking capabilities, and it has been used with its data logging capabilities to compare and record background, non-compliant areas, and complaint areas.

For comparative data, ppbRAE users suggest that once it has been calibrated, do not turn the instrument off until the investigation has been completed. The investigator should note the time each time the instrument changes locations (or change the site location on the equipment each time there is a location change) and start tracking by the following approach:

- Outside: 10 to 15 minutes
- Complaint room(s) indoors: 10 to 15 minutes
- Non-complaint room(s) indoors: 10 to 15 minutes

Figure 8.12 Example of Screening Survey Using a ppbRAE. Use for mold VOC detection in wall spaces, none destructive testing. (Courtesy of RAE Systems, Sunnyvale, CA.)

In some instances, some investigators have set a screening limit of 1,000 to 2,000 ppb (or 1 to 2 ppm) above the documented background levels. This does not necessarily mean the VOCs are a problem, but the investigator may want to investigate further. The present consensus is that one will typically get around 200 to 350 ppb indoors and 50 to 150 outdoors, and perfumes, cleansers, and air deodorants will greatly affect these readings.

Once an organic compound has been identified and the source is unknown, the investigator could use correction factors (i.e., relative instrument response) to determine actual levels and quantitate actual levels while "sniffing" out the source of the chemical(s). In one instance, a high school complaint with a follow-up investigation led to the identification of an ever-clear punch.

Presently, the ppbRAE is just another tool for indoor air quality investigators. The extent to which it may be used has yet to be fully determined.

Helpful Hints

As they are quite expensive, both the FID and PID can be and should be rented prior to making a purchase. If using an instrument for the first time, the investigator should also plan for an extra day of rental prior to using the equipment and spend time getting familiar with its operation as well as the data management procedures. A testing period is strongly advised prior to field use.

In one instance, a field technician was sent out on a job with a direct reading instrument. His supervisor thought, "Nothing could possibly go wrong. The instrument is dummy proof." The technician probed with the instrument into a hole in the wall, showed panic on his face, and rapidly departed the building. All other occupants who observed this event followed in fast pursuit. Upon further investigation, there was no real problem.

Make sure all instruments have been properly calibrated. Calibration is "confidence in a bottle."

SUMMARY

Volatile organic compounds are ubiquitous. They are outdoors in the country, and they are inside buildings. Their presence in indoor air quality is not so much an issue as the type of compounds, exposure levels, and duration of exposure.

Indoor air quality generally consists of the outside air and all the other complex contributions from within a building. Although they do not generally exceed workplace exposure limits, indoor exposures are more complex, and it has been suggested that these low level complex mixtures can cause a number of health problems similar to typical indoor air quality complaints.

Screening for volatile organic compounds is recommended in cases where chemicals are suspect only. If the source is known to be volatile organic compounds, more indepth sampling should be performed. Depending on the situation, sampling can be performed for knowns on the basis of chemical type (e.g., alcohols) or for unknowns. There are several different tools for accomplishing this end. Each has different applications and should be considered on a case-by-case scenario.

Remember, volatile organic compounds are not inclusive of all chemicals. Although one may be able to "suck air" into a container and take the container to a laboratory for analysis, the laboratory capablities are limited. If there are no significant findings by the means provided herein, the investigator can only state "no significant findings."

REFERENCES

1. Kennedy, Eugene, Ph.D. and Yvonne T.G. Evaluation of Sampling and Analysis Methodology for the Determination of Selected Volatile Organic

Compounds in Indoor Air. (Research document) NIOSH, Cincinnati, Ohio. December 1993.

2. ASHRAE. WHO Working Group Consensus of Concern About Indoor Air Pollutants at 1984 Levels of Knowledge. (Standard) Table C-4. ASHRAE Standard 62-1989. p.21.
3. Indoor Air Quality Management Group. Guidance Notes for the Management of Indoor Air Quality in Offices and Public Places. Government of Hong Kong Special Administration Region. November 1999.
4. Hodgson, Michael, MD, MPH, H. Levin, B. Arch, and P. Wolkoff, Ph.D. Volatile Organic Compounds and Indoor Air. Journal of Allergy and Clinical Immunology. 2(2):296-303 (1994).
5. Molhave, L. R. Bach and O. Peterson. Human Reactions to Low Concentrations of Volatile Organic Compounds. Environmental International. 12:167-75 (1986).
6. Grote, Ardith A. Screening Applications Using Thermal Desorption Techniques. Presented at the AIHC Exposition in Kansas City, Missouri on 23 May 1995. NIOSH, Cincinnati, Ohio.
7. Ness, Shirley A. *Air Monitoring for Toxic Exposures.* Van Nostrand Reinhold, New York, 1991. p. 59.

Chapter 9

MOLD VOLATILE ORGANIC COMPOUNDS & MOLD DETECTION

The Year 2000 started with a bang in that molds took center-stage in the overall consideration of indoor air quality concerns. Yet, sampling methodologies are expensive, difficult to interpret, and require long laboratory turnaround times.

There is controversy amongst the experienced environmental professionals. One investigator condemns a building, requiring extensive remediation with minimal sampling while another investigator attempts to assess similar scenarios by more extensive sampling and using more tools to gather additional information. One of these tools which is gaining in popularity is air sampling for mold volatile organic compounds.

Some investigators choose not to perform air sampling for the physical presence of molds in the air. Due to the lack of well-substantiated data and clear guidance for interpretation, many investigators simply inspect for evidence of molds. These inspections may be performed by means of visual observations, moisture testing, and/or odor tracking.

The purpose of this chapter is to provide a few more tools and approaches that the investigator may find useful under different situations. Some of these techniques may prove to be indispensable to some, spirit incantations and ghost busting to others. They are mere tools, techniques that can be very effective if used properly.

HEALTH EFFECTS AND OCCURRENCES

Some researchers feel that MVOCs may cause health problems. Whereas the scientists tend to focus on sensory irritation similar to that of VOCs, non-research investigators report a concern that the MVOC may cause headaches, eye and respiratory tract irritation, and dizziness. Presently, there is no substantiated data that correlates MVOC exposure levels to speculated health effects.

Metabolic by-products of molds are referred to as mold volatile organic compounds, or MVOCs, and consist predominantly of alkanes, alcohols, and ketones. The specific compounds that have been identified are dependent upon mold type (i.e., genus and species), food consumed, and environmental factors (e.g., moisture availability). At the present, researchers are scrambling to identify MVOCs common to all fungi and to determine variances between genera and species. See Table 9.1 for a list of some of the more consistently reported organics that may be anticipated wherever molds are encountered indoors. According to one laboratory that has performed over 600 analyses for various clients, the most frequently identified mold by-products are 3-methyl-1-butanol, 2-octen-1-ol, and 2-heptanone.[1]

Table 9.1 Listing of Mold Volatile Organic Compounds [2,3,4,5,6]

Compound	Characteristic Odor
1-Octen-3-ol [2,4,5,6]	musty, mushroom-like
2-Methyl-1-butanol [3,4]	
2-Methyl-2-butanol [6]	
2-Methyl-1-propanol [3,4]	
2-Octen-1-ol [4,5,6]	weedy
2-Pentanol [4,5,6]	
3-Methyl-1-butanol [3,6]	
3-Methyl-2-butanol [4,5,6]	
3-Methylfuran [3,4,5,6]	
3-Octanol [2,4,5,6]	nutty
3-Octanone [2,5,6]	sweet ester, metallic-like
1-Butanol [4]	
Dimethyl sulfide [3,4]	
Geosmin (terpene) [3,4,5]	earthy
2-Heptanone [4,5,6]	
2-Hexanone [4,5,6]	
2-methyl-isoborneol [5,6]	
2-isopropyl-3-methoxypyrazine [5,6]	

Some researchers are feel that MVOC sampling data may result in more reliable mold information than viable air sampling. They state that in visually moldy interior environments, air sampling sometimes yields "false"[7] negatives. These false negatives are often confirmed false where molds are found hidden in the wall cavities, under sheet vinyl, and within hidden spaces (e.g., behind sinks and cabinets). The hidden molds can be detected by odor and confirmed by MVOC sampling.

It is unclear as to whether the MVOCs are clearly causing health problems or whether the presence of MVOCs is merely an indicator. There have been no studies published which link the chemicals identified as MVOCs at the reported levels to be associated with health complaints. Reported levels have been around 50.5 $\mu g/m^3$ total MVOCs, 16.1$\mu g/m^3$ 2-octen-1-ol, and 1.8 $\mu g/m^3$ methylfuran. There are no ACGIH limits for either 2-octenol or methylfuran, and they are considered slightly toxic and moderately toxic, respectively.[8] The other compounds listed as MVOCs are slightly to moderately toxic.

On the other hand, however, investigators have used MVOC sampling not only to locate but to rule out the presence of molds in wall spaces. In one case, a consultant recommended an entire residence be leveled to the ground. This was based on known water damage, some visible molds, and minimal sampling. Another consultant investigated, drilled a small hole in each wall, took multiple samples, and isolated the problem area to a couple walls. The end result was remediation of a couple walls and some wall sections. The occupants returned to the residence, and their health complaints did not recur.

Researchers are attempting to determine the efficacy of identifying mold genus and species by MVOC fingerprinting. Although there has been some consistency in results with distinct differences between species, most research has been performed under controlled laboratory conditions with well-defined nutrient agars. Within the same species (e.g., *Penicillium varotii*), many of the MVOCs grown on malt extract agar are different from those grown on dichloran glycerol agar. In a building, the nutrients are variable. There are likely to be more than one type of mold contributing to the total indoor MVOC, and MVOCs that are detectable are typically lower than indoor background VOCs. Whereas the indoor VOCs associated with off-gassing of building materials and furnishings may range from 200 ppb to 2 ppm, high levels of total MVOCs may range from 0.01 $\mu g/m^3$ to 2,000 $\mu g/m^3$, averaging around 33 $\mu g/m^3$ (i.e., 8 ppb as compared with hexanone).[9] For MVOC findings from composites reported by one laboratory, see Table 9.2.

To further complicate matters, bacteria produce VOCs as well. There has been limited research regarding bacterial VOCs, yet bacteria can contribute to the MVOCs or be the primary contributor. Bacterial VOCs may potentially muddy the waters when interpreting these common VOCs with the assumption that they are clearly mold by-products. The predominant by-products produced by both bacteria and molds are 1-propanol, 2-butanol, and dimethyl trisulfide.[11] It should also be noted

Table 9.2 Microbial Volatile Organic Compound Database[10]

Location/Types of MVOC	Average Concentration ($\mu g/m^3$)	# of Positive Sites of Total Sites Sampled
Indoor air samples		
Total MVOC	50.5	106 of 119
1-Octen-3-ol	4.8	40 of 119
2-Octen-1-ol	16.1	58 of 119
3-Methylfuran	1.8	4 of 119
Outdoor air samples		
Total MVOC	6.5	8 of 20
1-Octen-3-ol	1.5	2 of 20
2-Octen-1-ol	3.0	3 of 20
3-Methylfuran	N.D.	0 of 20

that actinomycetes (e.g., 3-methyl-1-butanol, dimethyl trisulfide, and geosmin), algae, and trees (e.g., terpenes) may produce similar VOCs as well.[11,12]

SAMPLING FOR MVOCs

Sampling for MVOCs is an evolving science. There is little or no published guidance available from government sources, and there are varying opinions amongst laboratories. The information provided herein is the most recent information available prior to publication of this book. Locate a laboratory and discuss their experience and capabilities for analyzing MVOCs. The protocols will remain in flux until NIOSH or another government entity publish their own method.

Sampling Strategy

Careful consideration should be given to where sample data will provide data that the investigator can best interpret. Exposure sampling may be performed in order to determine exposure levels and assess the MVOC impact on occupant health. These samples should be taken within occupied areas.

Diagnostic sampling may be performed in order to locate areas in wall spaces where destructive sampling is not desired and the investigator is attempting to locate where molds are actively growing. These samples should be taken within the wall space(s) or other inaccessible areas

Once the decision has been made as to whether to sample for occupant exposures or wall spaces, the investigator may wish to target sites that are likely to represent worse case scenarios. This may be done through odor tracking, comments from building occupants, or low-level VOC detection instrumentation (e.g., ppbRAE). The latter approach shows promise as a means for tracking MVOCs to the source. The author's experience has been that of tracking molds using a ppbRAE like a tracking device. As the site where molds are growing is approached, the readings on the meter become elevated. Where background has typically been less than 300 ppb, VOC levels at identified moldy areas may exceed 1,000 ppb, sometimes going as high as 20,000 ppb (or 20 ppm).

Comparative samples should also be taken with a minimum of one outside sample. If possible, an indoor non-complaint or non-problem area should be sampled as well when performing occupied space sampling. This is particularly important where there are no guidelines for interpreting results.

Sampling Methodology

Air sampling for MVOCs has been and can be performed by the same methodologies as those used for VOC. More recently, however, specific mold organic

Figure 9.1 Air Sampling Pump, Anasorb® Tubes, and Cooler for Shipping

by-products have been targeted, and an MVOC method developed for those organics that have been commonly encountered in various studies.

Although the specific MVOCs that are targeted may vary by laboratory, the end result is exclusion of the VOCs associated only with building materials, furnishings, and activities. Otherwise, you may not "see the forest through the trees." The MVOC target compound sampling technique is more focused than the broad VOC methods and easier to interpret.

The method requires the following:[5]

- Air sampling pump
- Capture medium: SKC Anasorb® 747
- Flow rate: 0.2 liters per minute
- Range air volume: 48-120 liters
- Desorbtion: dichloromethane
- Standard: target compounds
- Analysis: GC-MS
- Limit of detection (each compound):
  0.1-0.5 μg/1 ml injected dichlormethane
- Special handling: cap and freeze media for shipping and storage

Samples may be taken from the occupied air within a building or from within a wall space/area suspect of having MVOCs.

SCREENING METHODOLOGIES

A building inspection for molds is one of many diagnostic tools that may be used for assessing a building for potential mold growth. They go outside "sampling" envelop and enter into other disciplines (e.g., building construction technology). At the same time, these techniques support each other and are most effectively simultaneously. The investigator should be open to all possibilities.

Visual Observations

As discussed in previous chapters molds have known, predictable habits and habitats. They thrive in high moisture. They require organic material for nutrition, and they do not require light.

Most molds require in excess of 80 percent moisture on and/or in a food substrate. See Tables 5.3 and 5.4 for moisture requirements of some molds. In buildings, the food substrate may include, but not be limited to, wet gypsum board paper, damp carpeting, wet wood flooring and trim, damp upholstery, moist areas under vinyl wallpaper and floor tiles, wet ceiling tiles, and evaporation in duct insulation.

Moisture may be due to a plumbing pipe/fixture leak, sewage leak, building structural leaks (i.e., window jams, doors, and roofs), moisture buildup from uncured concrete foundation, moisture infusion due to temperature differentials in a building, and condensation in the air handling system.

Plumbing leaks are the easiest, most commonly identified source of moisture buildup. The most common areas are under kitchen and bathroom sinks, in and behind cabinets. The more difficult plumbing leaks to detect are those around bathtubs and toilets. Second story leaks tend to show wherein water stains become evident or water comes out from around wall penetrations (e.g., air supply louvers and lights). Even more difficult to locate are damaged pipes in wall spaces. Moisture buildup inside a wall that has vinyl wallpaper can be difficult to locate as well. Yet, the glue for most wallpaper is water-soluble, and moisture buildup will loosen surface adhesion, making the wallpaper easy to lift. If there is no wallpaper, a water spot may or may not appear, depending on the extent of the leak.

Water and sewage leaks may occur around toilets or kitchen disposal units or a cracked concrete slab. Foundation shifts provide a conduit for broken sewage pipe moisture and gases to return to the interior air spaces.

Improperly sealed door and window jams are typically a means whereby moisture may enter the indoor air spaces, particularly where there is considerable moisture (e.g., excessive rain). Extreme temperature differences between the indoor air and outside air (or indoor air and ground) may result in moisture intrusion.

Roof leaks are a common source of moisture intrusion as well. These are a little more difficult to track. Even if there are water stains on a ceiling, the actual site of a leak may not be easily traced due to extensive migration within the roof space(s).

Concrete foundations require 28 days to cure, but many contractors do not allow more than a couple days for curing. Moisture is created in the curing process, and when resilient floor tiles or vinyl sheeting are laid over the uncured surface, moisture becomes trapped and accumulates. It should also be noted that the old asbestos floor tiles permitted moisture to escape, and the tiles were typically laid down with a solvent-based glue. The non-asbestos tiles trap the moisture, and these tiles are typically laid down with a water-based glue. When the water-based glues get wet, the glue begins to fail, and the tiles begin to lift. Commercial carpeting may also have been put down with water-based glues. Thus, poor adhesion of floor surface materials is a good indicator of potential moisture buildup.

Condensation in a commercial air handling system and window air conditioners occurs at the cooling coils. In large commercial units, the condensate is collected in a drip pan where it should drain out of the unit. Often the unit fails to drain the full amount of settled water. Sometimes the drain gets plugged. This condensate build up in the drip pan is a frequent source of moisture for molds to grow. Yet, it does not stop here.

The moisture from the coils and drip pan is blown onto the surface immediately after the cooling coils. This surface may or may not be insulated. The author has observed molds growing on what would otherwise appear to be a clean metal surface. More frequently, however, this area immediately after the cooling coils is covered with impregnated fiberglass insulation. A felt-like surface covering may be observed which may later to be confirmed as mold.

Poorly insulated duct or non-insulated spots in walls may also be a source for condensation buildup, especially where the temperature difference is extreme between the air in the ducting and outside in an attic space. The condition of the duct and insulation may observed and moisture buildup can be observed. Sometimes condensation occurs at the air supply registers. Molds may be observed growing on the outer edge of the louvers or just inside.

The nutritional source may be glue, paper, decayed organic matter, organic debris, soil, and human food. Many of these materials are found in house dust that can be found in all areas mentioned above.

Odor Tracking

Odor thresholds for some of those by-products produced by molds are quite low. For instance, the odor threshold for dimethyl trisulfide has been reported as low as 10 ppt. For 1,5-octadien-3-ol, the threshold has been published as low as 1 ppt, and geosmin is noticed at 5 ppt.[13,14] For comparison, the very distinct odor for hydrogen sulfide is noticeable for many at 4.5 ppb, and for a limited number of people, the rotten egg odor is sometimes identified at levels as low as 70 ppt.

Note that there are no ACGIH or EPA standards for dimethyl trisulfide, 1,5-octadien-3-ol, or geosmin whereas the ACGIH TLV-TWA for hydrogen sulfide is 5 ppm.

Thus, in the case of MVOCs, odor should not be mistaken as an indication of excessively high levels of MVOCs or health effects.

To overstate the obvious, some people have a greater sense of smell than others. Those of such great fortune (or misfortune) who are able to track odors may go to the source. Sometimes building occupants can direct the investigator to a source.

Odors are typically associated with visually observed molds out in clear view, but sometimes odors can be isolated and molds located behind or in hidden spaces. These hidden spaces may include, but not be limited to, wall spaces (e.g., behind light sockets), around windows, areas under carpeting, behind bathroom sinks, and ceiling spaces.

One drawback to tracking odors is that by the time a moldy odor is noticed, the odor has diffused into the general air space and/or been mixed with the air from within the air handling unit. For this reason, if the source is the mechanical air ducts, interpretation can be particularly difficult. Odor alone will rarely isolate the problem. Other methods must be used.

Moisture Testing

Where moisture cannot be observed and is suspect, a moisture detector is quite useful. Without a moisture meter, an investigator may overlook potential areas for mold growth, or suspect areas must be demolished. The later culminates in poor aesthetics, potential release of enclosed mold spores, and expensive repairs.

A moisture meter detects moisture through paint, varnish, wallpaper, floor tiles, wood, and carpeting. There are non-destruct (e.g., Tramex® Moisture Encounter) and probe instruments (e.g., Tramex® Probe Encounter). The former is that which is used most frequently in indoor air quality investigations.

The Moisture Encounter detects moisture beneath a surface by measuring the resistance between two low frequency signals transmitted from conductive pads on the base of the instrument. When resistance is decreased by the presence of moisture, the needle on the meter deflects accordingly and reads in relative percent.

It should be noted that metal will also defect the meter. If the material being tested has metal composition or structural steel behind the surface, the meter will read high moisture content. Some gypsum board has an aluminum foil backing to provide thermal insulation where the gypsum board is used along exterior walls. The aluminum foil will deflect the meter full-scale, result in erroneous readings regarding moisture content.

The Moisture Encounter has three scales that are based on penetration and sensitivity, and it has two ranges. The relative range is that used most frequently. The wood and timber setting, Scale 1, penetrates ½ to ¾ inch deep. Indoor wood should read less than 10 percent on the % $H_2 0$ Wood indicator range.

The drywall/roofing setting, Scale 2, penetrates 1 inch deep, and readings are taken on the relative range of the meter. This is the most sensitive setting and most frequently used scale in indoor air quality investigations. As the meter may be used to penetrate carpeting, floor tiles, and roofing shingles, Scale 2 may be used to locate moisture hidden below surfaces. The origin may be the underlying concrete, trapped moisture between floor and wall surfaces, or roof leaks. In regards to roof leaks, the meter may be used to pinpoint the source area. If there is wood flooring, Scale 2 can penetrate the surface sufficiently to determine moisture that lies beneath after Scale 1 has determined low moisture content in the wood.

The plaster/brick setting, Scale 3, penetrates ¼ to ½ inch deep, and readings are taken on the relative range. This is the gypsum (also referred to as drywall) setting, and it is the least sensitive scale. On Scale 3, the most frequently encountered moisture problem identified in walls is associated with moisture trapped behind vinyl wallpaper, moisture rising by capillary movement from the ground up within walls, and moisture movement from a leak. Moisture and mold may not be visually observed under wallpaper and painted surfaces. Thus, the meter permits a non-destructive assessment for potential molds associated with moisture.

Information regarding the penetration of each setting can be used to track depth of moisture-containing material. For instance, on Scale 3, a vinyl wallpaper-covered wall may read zero, but on Scale 2, the same wall may read 95 percent relative. The different readings can be interpreted to mean the moisture is not immediately below the wallpaper. It is as much as 1 inch deep. In most cases, this would mean the gypsum board may have moisture on the inside away from the wallpaper. The term "may" is a caveat that depends on other factors, such as no metal components within the wall space evaluated.

Figure 9.2 Tramex® Moisture Encounter and Moisture Meter with Probe-The Moisture Encounter is being used to test for moisture under the floor tile (left), and the Moisture Meter is being used to probe wood siding on a building.

As the clarity and rationale of Scale 2 being the most sensitive and Scale 3 the least is most allusive, the investigator should be clear on the penetration capabilities of each of the scales. It may be helpful to write the penetration capabilities on the instrument.

INTERPRETATION OF RESULTS

When assessing occupied space MVOCs, indoor sample results should be compared to outside samples. Short of concluding that the results outside are a given multiple of that indoors, identified MVOCs that have exposure limits may be compared with other standards in a similar fashion to that of VOCs. It is highly unlikely, however, that any known standards will be exceeded. Otherwise, the investigator should seek the assistance of someone with more experience, expertise.

When assessing wall space MVOCs, in-wall sample results should be compared to outside samples and possibly samples obtained from wall spaces known not to have or to be associated with mold growth. Sample results will, once again, be relative to known non-problem areas.

Positive findings upon the visual observation of molds should be confirmed. That which appears to be mold is not always such. For instance, one building owner was prepared to close down a restaurant and spend large sums of money on remediating a black velvet-looking spot in the kitchen area. Confirmatory sampling disclosed that the mold was indeed grease with carbon particles from cooking.

On the other hand, negative visual observations do not necessarily mean there are no molds. Moldy, earthy, musty odors may be traced to enclosed, hidden areas, and a moisture meter may identify high moisture areas. Both scenarios are likely to be associated with molds. In these cases, further investigation by demolition, surface sampling, and/or mold air sampling is indicated.

Even when MVOC sampling and a building inspection fail to identify any observable molds or molds that can be found through demolition, air sampling may still indicate a problem. Then, the onus comes back to the inspector. The search goes on!

SUMMARY

MVOC sampling and building inspections have been added to the investigator's bag-of-tricks. MVOC sampling is best utilized to locate mold growth in hidden spaces. Visual observations can be very effective in identifying probable sources of mold growth, and the more knowledge an investigator has regarding building construction technology, the more effective a building investigation.

Characteristic mold odors and MVOCs may be traced to a source or site(s) where there is potential mold growth. A moisture meter may be used to locate moisture, growth havens for molds. The combined approach for identifying growth sites can be highly effective. Yet, a visual inspection of suspect areas may be required for confirmation.

REFERENCES

1. Air Quality Services. Using Microbial Volatile Organic Compound Analysis to Detect Mold in Buildings. *AirfAQS Extra.* 5(3).
2. Burge, Harriet A. *Bioaerosols.* CRC Press, Boca Raton, Florida. (1995). p.258.
3. Sunesson, Anna-Lena, et.al. Identification of volatile metabolites from five fungal species cultivated on two media. *Applied and Environmental Microbiology.* 61(8):2911-18 (1995).
4. Wessen, Bengt and K. Schoeps. Microbial Volatile Organic Compounds-What substances can be bound in sick buildings? *Analyst.* 121:1203-05 (1996).
5. ACGIH. *Bioaerosols: Assessment and Control.* ACGIH, Cincinnati, Ohio (1999). p. 26-1.
6. Aerotech Labs. IAQ Tech Tip #42: MVOCs. www.aerotechlabs.com (October 2000).
7. Conference Venue: Natcher Conference Center at National Institutes of Health, Bethesda, MD. 1:245 (Sept. 27-Oct. 2,1997).
8. Lewis, Richard J. *Sax's Dangerous Properties of Industrial Materials.* John Wiley & Sons, Inc. New York, New York. 10th Edition (2000).
9. Air Quality Services. Using Microbial Volatile Organic Compound Analysis to Detect Mold in Buildings. *AirfAQS Extra.* 5(3).
10. Moray, Phil, A. Worthan, et. al. Microbial VOCs in Moisture Damaged Buildings. *Healthy Buildings/IAQ 97.* Conference Venue: Natcher Conference Center at National Institutes of Health, Bethesda, MD. 1:247 (Sept. 27-Oct. 2,1997).
11. Burge, Harriet A. *Bioaerosols.* CRC Press, Boca Raton, Florida. (1995). p.260.
12. Burge, Harriet A. *Bioaerosols.* CRC Press, Boca Raton, Florida. (1995). p.250 & 258.
13. Burge, Harriet A. *Bioaerosols.* CRC Press, Boca Raton, Florida. (1995). p.251.
14. Pengfei, Gao. MVOCs. E-mail (September 28, 2000 at 5:38 PM). Center for Disease Control.

Chapter 10

CARBON DIOXIDE

An angry complaint is lodged, "There is no oxygen in this room. Everyone is passing out!" The comment comes from an office occupant in a building where this health complaint is unique. The occupant had previously complained about the cold air supplied in her office space. Subsequently, her air supply was turned off, and when accommodating visitors, she closed the door to her already stuffy office. The predictable results were elevated levels of carbon dioxide in a confined space.

Ever present in our outdoor environment, carbon dioxide levels are normally higher indoors than outside. The elevated levels indoors may be due to combustion, leaking compressed gases, and animal respiratory by-products. In indoor air quality, carbon dioxide is primarily an indicator gas. It is an indicator of inadequate fresh air or insufficient dilution of air contaminants generated in a building.

The source of complaints may be formaldehyde off-gassing from furniture, but the elevated levels of formaldehyde may be alleviated with an increase in fresh air entering the building. The carbon dioxide levels provide information as to the adequacy of fresh air supplied to the occupied spaces. This is the first consideration in rectifying an indoor air quality concern.

Occasionally, however, carbon dioxide contributes to health complaints in a subtle fashion. In some cases, indoor air quality investigations are associated with elevated carbon dioxide levels with complaints of stuffiness and inadequate air. The sense of inadequate fresh air may be voiced in terms of "can't breath" or a "stuffy, suffocating sensation."

Rarely are the carbon monoxide levels excessive to the point of causing a significant impact on human health. At levels well in excess of those normally found in indoor air quality studies, carbon dioxide is a simple asphyxiant. When levels approach percent of total air, carbon dioxide begins to displace the much-needed oxygen. The most sensitive sign of reduced oxygen in the breathing air is a reduced ability to detect slight differences in the brightness of objects. As the levels increase, the symptoms progress from an increase in heart and/or respiratory rate to headache and a feeling of fatigue. In extreme intoxication, exposed individuals may experience nausea and vomiting, progressing to collapse and unconsciousness.

As an indicator, carbon monoxide levels are used predominantly to determine adequacy of the air exchange and to isolate stagnant air pockets where there is little or no air movement. Although the air exchange in an office building may be adequate (e.g., fresh air intake calculated at 20 CFM/person), areas within the same building may not receive their share of the total supplied air to the building. Their air supply vents may be turned off or provide, at best, diminished flow due to reworked duct systems during renovation. The latter is common wherein there are constant add-ons and changes to the occupied spaces.

For those rare occasions whereby carbon dioxide levels may become excessive, the investigator should be aware of the potential source buildup and cause for the increased levels. The assessment of carbon dioxide levels is a basic tool to be utilized in all indoor air quality investigations.

OCCURRENCE OF CARBON DIOXIDE

Carbon dioxide is a colorless, odorless gas. When mixed with water, it is referred to as carbonic acid that has a slightly acid taste.

Humans and other animals inhale oxygen and expel carbon dioxide as a waste product. Plants inhale carbon dioxide and expel oxygen. Over time the balance of available oxygen and carbon dioxide has shifted in the ambient environment. Although reported as low as 0.034 percent (i.e., 340 ppm), ambient levels have been elevating slightly over time to around 0.037 percent (i.e., 370 ppm). There are slight variations in these levels, depending on the time of year (e.g., plant uptake of carbon dioxide occurs during growing season in the summer), weather (e.g., air inversions may trap and result in an increase of all air pollutants), and industrial exhaust (e.g., combustion by-products).

In indoor air quality, the primary source of carbon dioxide is human expelled air. The expelled air builds up in airtight buildings, confined air spaces (e.g., enclosed offices with no air supply), overcrowded spaces (e.g., classrooms), and high activity areas (e.g., health clubs).

Other sources of carbon dioxide are by-products of combustion, sugar fermentation, and ammonia production as well as carbonated beverages, compressed carbon dioxide (e.g., fire extinguishers), dry ice, and aerosol propellants. Carbon dioxide may be evolved indoors by gas cooking appliances, space heaters, wood-burning appliances, and tobacco smoke. Industrial exhausts whereby burning is involved (e.g., sanitary waste) should not be ruled out. If potential contributors to the total carbon dioxide levels, these sources should be considered in the investigation. They may require a special local exhaust or further evaluation of an exhaust generated by an industrial process and captured by the air intake for the building under investigation. In the latter case, the finding may provide information leading to the actual source of the problem (e.g., organic chemicals in the combustion by-products from an industrial operation). In the latter case, closure of the fresh air intake may be the solution.

In accordance with ASHRAE, levels in excess of 1,000 ppm should be considered an indicator of insufficient make-up air into a building. The recommended limit to minimize health effects according to ACGIH is 5,000 ppm.

SAMPLING STRATEGY

In offices and conditioned school buildings, the investigator should perform air monitoring for carbon dioxide to assess the adequacy of the make-up air. This approach should become part-and-parcel of one's standard toolbox. Not only should various areas throughout a building be assessed, but outside air sampling should be performed as well.

Consider the conditions, and determine sample locations accordingly. Conditions to be considered should include, but not be limited to, the following:

- Different air handling zones
- Worse case complaint areas
- Non-complaint areas
- Heavily occupied areas during peak occupancy periods
- Enclosed office spaces
- Center of cubicle areas
- Occupied and non-occupied spaces
- Different activity areas
- Time of day and occupancy

In areas where the occupancy and activities are not consistent, the carbon dioxide levels will fluctuate accordingly. In a typical office building, the carbon dioxide levels will be close to the outdoor levels at the beginning of the day and peak out between 1 PM and 2 PM in the afternoon. A thorough assessment of each site would involve a minimum morning, noon, and late afternoon sample or placement of a data logging instrument at one site, leaving it at the same location all day. The latter scenarios are, however, unfeasible in most cases involving multiple sampling. In such cases, it would be advisable for the investigator to choose one or two locations to assess for changes throughout the day.

In residences and unconditioned schools, the investigator may perform air monitoring for carbon dioxide to assess air tightness of the structure. Where windows are open, the point is forgone conclusion. If, however, inside air monitoring is accomplished, outside air sampling should be performed as well.

The investigator may choose to take numerous samples and contour an area. This can be informative while providing an easy to review visual of recorded numbers and relative references. The number of samples collected, however, is limited by method of air sampling.

SAMPLING METHODOLOGIES

Air sampling may be performed by direct reading instrumentation or colorimetric detector. Both methods provide instantaneous data, and the direct reading instrumentation provides on-going data. Direct reading instruments incur a high initial cost, and colorimetric detector tubes involve a low cost per sample. The choice of method should be based on desired number of samples and accessibility of equipment.

Direct Reading Instruments

If a large number of data points are to be collected and equipment is readily available, direct reading instrumentation is the most feasible approach. They are available both for purchase and for rental. For low usage and minimal maintenance situations, rental is the most practical means for accessing equipment.

There are several direct reading instruments that are designed to measure carbon dioxide levels. Many of them are referred as "indoor air quality monitors."

Carbon dioxide monitoring equipment will generally be part of a four-channel monitor, some that provide information on typical concerns of indoor air quality (e.g., temperature, relative humidity, carbon monoxide, and carbon dioxide), and others that provide four gases chosen at the discretion of the investigator (e.g., carbon monoxide, carbon dioxide, hydrogen sulfide, and total volatile organic compounds).

Most are data logging instruments that record data throughout the monitoring period that can latter be retrieved. Data retrieve is convenient and makes time-exposure data readily accessible for review. See Figure 10.1

Figure 10.1 Indoor Air Quality Monitor–Records and data logs carbon dioxide along with temperature, relative humidity, and carbon monoxide. (Courtesy of KD Engineering in Bellingham, WA)

Colorimetric Detectors

If a small number of data points are to be collected and direct reading equipment is not readily available, colorimetric detection is the most feasible approach. The initial purchase of a pump and detector tubes is relatively low in cost, and equipment is easy to carry. However, when the sample numbers exceed ten, the cost of the tubes starts to outweigh the cost of direct reading equipment rental.

The term "detector" is an operative term in that colorimetric detectors detect levels instead of providing precise data. Yet, they are simple, easy-to-use, and practical for spot checks and collecting limited samples when direct reading instrumentation is not available.

Colorimetric detection of chemicals is vintage science that has been available for over fifty years. Detection is based on a chemical reaction between the specially prepared sample tube and the gas or vapor.

Sampling equipment includes a hand held bellows or piston pump and detector tubes, specific for the chemical in question (e.g., carbon dioxide). The pump and tubes must be from the same manufacturer. Do not mix products (e.g., a Drager® tube with an MSA pump). The tubes are chemical specific with different ranges (e.g., 0.01 to 0.3 percent by volume which is 100 to 3,000 ppm).

The detection methods have a manufacturer-rated accuracy of 15 to 25 percent, depending on the manufacturer. Some experienced investigators, however, have found variances in the low range that depart from direct reading instrument data taken at the same location by as much as 50 percent.

When the components within the tube react with the gas or vapor being sampled, there is a color change (e.g., white to violet). The tube is marked with generalized detection levels, and the length of the color change provides information regarding the detection level at that point in time. See Figure 10.2.

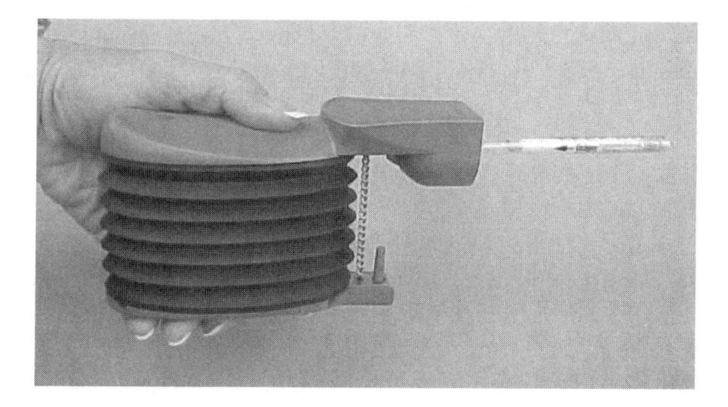

Figure 10.2 Drager® Pump and Colorimetric Detector Tube for Carbon Dioxide

Determine potential interferences and/or cross sensitivities. There is no cross sensitivities listed for Drager® carbon dioxide detector tubes.

The sampling method is summarized as follows:

- Equipment: bellows or piston pump
- Sampling Medium: carbon dioxide detector tubes in desired range
- Procedure: insert tube in pump holder with arrow pointing toward the pump, squeeze and release the pump for the number of strokes designed by the manufacturer (1 to 10 times), allowing the pump bellows or piston to fill completely between each stroke
- Analysis: read the concentration on the basis of the perceived length of the color change

HELPFUL HINTS

Take prolific notes, and record sample locations on a building schematic. The best reference is a visual with recorded instantaneous data or long-term averaged data. If it is not possible to record the sample locations on a schematic, document the precise sample location and rationale for choosing each sample location (e.g., no air movement). Always record the time each data point was collected.

Don't hold or place the monitoring probe or sample inlet close to your face. Breathing into the monitor will result in unreasonably elevated data. The placement of a monitor for long-term data logging should also be placed as to avoid other people from breathing directly into the monitor probe. Upon final analysis, however, a well-planned sample site may culminate in a few surprises (e.g., a small mid-day meeting of people around the vicinity of the monitor). Attempt to keep track of activities, and never assume an unattended monitor will remain untouched.

When weighing an equipment rental verses a purchase, consider the cost of calibration time, span gases, and annual replacement costs for sensors against the cost of rental and freight charges. Whereas it may appear easier to purchase, the long-term maintenance costs could outweigh the cost of rental and freight.

INTERPRETATION OF RESULTS

Where an entire building has consistent levels of carbon dioxide in excess of the ASHRAE recommended limit of 1,000 ppm, interpretation is easy. There is insufficient fresh air. This scenario is, however, rare.

Assess the results in terms of relative time and space. Some of the measurements may exceed 1,000 ppm for only a few minutes. All areas may average 900 ppm throughout the day, or a single office space may have levels that 1,000 ppm.

Attempt to determine the cause of these slight variances. It may not always be due to insufficient outside air. The cause may be isolated to no air movement, insufficient air supply to an area, or unusually high traffic on the day of the monitoring.

Where the levels exceed the ACGIH recommended limit of 5,000 ppm, seek other sources besides human exhaled carbon dioxide. It may be coming from by-products of combustion, industrial exhausts, carbonated beverages, leaking compressed carbon dioxide containers, dry ice, tobacco smoke, and aerosol propellants.

SUMMARY

Carbon dioxide monitoring is primarily a diagnostic approach for determining the adequacy of the fresh air supplied to a building and/or occupied space. The source of elevated carbon dioxide levels is generally, but not always, humans. Other sources are possible, and limits are not precisely defined. Be open to all possibilities!

Chapter 11

CARBON MONOXIDE

According to the American Medical Association, one thousand, five hundred people die annually due to accidental carbon monoxide exposures. Over ten thousand people seek medical attention, and there are many more who experience the toxic effects that go unreported. Medical experts speculated the reason for under reporting is that the symptoms of carbon monoxide poisoning resemble other common ailments.

Carbon monoxide displaces oxygen in the oxygen to blood transfer in the lungs. It has an affinity for the oxygen-carrying sites on the hemoglobin in the blood of 210 to 240 times greater than oxygen. As it displaces the oxygen, carbon monoxide prevents the distribution of the needed oxygen. The tissues become oxygen starved, and the individual fails to receive adequate oxygen supply. The higher the carbon monoxide levels, the greater displacement of oxygen occurs, and the more oxygen deficient the individual becomes.

Initial symptoms may be mistaken for the common flu or a cold. This includes shortness of breath on mild exertion, mild headaches, listlessness, and nausea. As exposures increase, the individual may experience severe headaches, mental confusion, dizziness, nausea, rapid breathing, and fainting on mild exertion. In extreme cases, not normally encountered in indoor air quality situations, exposures may result in unconsciousness and death.[1]

In indoor air quality, the initial symptoms of carbon monoxide poisoning may be difficult to differentiate from other causes of health complaints. After determining the occupants have symptoms that may indicate carbon monoxide poisoning, the investigator should be prepared to identify potential sources and perform air monitoring.

OCCURRENCE OF CARBON MONOXIDE

Carbon monoxide is a colorless, odorless gas. It is a flammable, combustion by-product, an important contributor to heat production. Carbon monoxide provides more than two-thirds the heat value created by carbon-based, combustible materials. In oxygen-rich air, carbon monoxide burns to form carbon dioxide. When there is insufficient oxygen, however, carbon monoxide does not burn completely and becomes one of the many by-products of combustion.

Incomplete combustion by-product of carbon-based material consists of other by-products as well. These products often have a color (e.g., red nitrogen dioxide gas) or an odor (e.g., aldehydes and volatile organic compounds). Although carbon monoxide is colorless and odorless, incomplete combustion by-products may be detected. If one complains of gasoline smells, the presence of volatile organic compounds should set off alarm bells to the investigator.

Combustible carbon-based materials include wood, charcoal, coal, natural gas, gasoline, diesel fuel, kerosene, oil, organic waste, and tobacco products. Materials less commonly implicated in indoor air quality studies are building, automotive, furniture, interior, and decorative materials involved in fires. Although rare, the latter scenario involving fire damage in or around the complaint area should not be overlooked.

In most instances, sources of elevated levels of carbon monoxide and other combustion by-products are encountered in one, or a combination of, the following scenarios:

- Vehicle exhaust in a building (e.g., air intake to a building located at street level in a busy alley or an automobile left running in a garage)
- Poorly vented gas-fired hot water heaters
- Air from a leaking exhaust duct or furnace flue
- Combustion by-products vented close to an air intake for a building
- Poorly sealed wood-burning stoves
- An insufficient amount of oxygen supplied to a gas-operated space heater
- Poorly tuned forklift trucks
- Poorly vented and insufficient replacement air in a building with natural gas and wood burning fireplaces (e.g., tight buildings with little or no air coming from outside to replace the air discharged through the chimney)
- Gas-fired heaters in air handling units
- Industrial combustion gases from an associated building

Some people exposed at home or work to carbon monoxide may also be exposed in the automobile that has a leaky muffler with the exhaust entering the interior of the car, or they may receive significant exposures when smoking tobacco products. Other building occupants may have contributing levels from other sources not directly associated with the building undergoing investigation. Be prepared to seek other source exposures where individuals experience symptoms indicative of elevated carbon monoxide levels, and the levels monitored during the building investigation are not elevated enough to cause the symptoms purported by a small portion of the occupants.

It is of note that carbon monoxide levels in the blood hemoglobin have a 5-hour half life. This means it takes 5 hours for half of the carbon monoxide tying up the oxygen receptor sites in the blood to dissipate when an individual is removed from the elevated levels of carbon monoxide. Recovery may be improved where a patient is placed on 100 percent oxygen.

The ACGIH recommended limit for carbon monoxide is 25 ppm averaged over 8 hours. The Canadian guideline for residential indoor air quality is 11 ppm for 8 hours and 25 ppm for 1 hour. The EPA National Emissions Standard is 9 ppm for 8 hours and 40 ppm for 1 hour. Background carbon monoxide levels in rural environments are typically zero.

SAMPLING STRATEGY

Occupants in an office space complained of severe headaches, nausea, and chronic fatigue that persisted into the weekend. The office space is located above industrial activities whereby propane forklift trucks are operating constantly. In the morning, diesel trucks back up to loading docks and remain running for the duration of the loading process. The exhaust plume from the trucks had been observed blowing toward the building eaves and occasionally into the air intake for the office spaces.

The evidence was overwhelming that carbon monoxide was the culprit, but air monitoring was met with resistance from the industrial leaser of the office space. As the sampling strategy turned into an intense chess game, the investigator should be aware that the obvious may become an unrivaled challenge.

The investigator should perform air monitoring for carbon monoxide whenever complaints include symptoms of carbon monoxide poisoning and a potential source has been identified. Even if an unidentified or industrial exhaust source is merely suspect, the investigator should perform air monitoring.

When the source has been identified, one worse-case sample may be taken for diagnostic (not exposure) monitoring. This would involve a sample at the source, not at the breathing zone of the occupants, the rationale being that if the levels are the source, by the time the carbon monoxide reaches occupants' breathing zones the exposures will be considerably diminished. Yet, in this case, to assure the source has been identified, one occupant exposure sample should be taken as well. This approach is a minimal.

As the time of occurrence may not be clear, ongoing monitoring should be performed for the duration of the building occupancy. In an office building, occupancy may occur from 7 AM to 6 PM. In a residence, occupancy may include 24 hours. In long term sampling, if the monitoring equipment is left unattended, the investigator should check prior to sampling to determine if the suspect event is likely to occur and confirm the occurrence of the event after sampling. For instance, if the suspect event is early morning diesel truck traffic and the trucks fail to deliver on a morning when they were scheduled, sampling should be rescheduled for a time to include truck deliveries.

If attempting to sample in a confined air space (e.g., closed conference room), confirm the doors remained closed during the sampling period. If you place a "do not disturb" sign on the door and on the monitor, do not assume compliance. Some additional measures may be needed to assure the area is secured.

In urban environments, always take an outside background air sample for comparison. Where the investigator monitors elevated levels indoors, outdoor air sampling should be performed. All contributing outdoor sources should be recorded.

SAMPLING METHODOLOGIES

Whereas direct reading instrumentation or colorimetric detectors both provide instantaneous data, the direct reading instrumentation provides on-going data. There are higher initial costs for instrumentation, but detector tubes involve a low cost per sample. The desired number of samples and accessibility of equipment are important considerations to choosing the more practical approach to air sampling.

Direct Reading Instrumentation

If a large number of data points are to be collected and equipment is readily available, direct reading instrumentation is the most feasible approach. They are available both for purchase and for rental. For low usage and minimal maintenance situations, rental is the most practical means for accessing equipment.

Carbon monoxide monitoring equipment may be purchased or rented as dedicated units or part of a four-gas monitor. Most are data-logging instruments that record time-exposure data. Some have a recorder that provides ongoing data that is recorded onto a strip chart while others simply provide a digital readout of data points that the investigator must record on paper.

Figure 11.1 Drager® Data-Logging Toxic Gas Monitor

Colorimetric Detectors

Colorimetric detection is the most feasible approach wherever a small number of data points are to be collected and direct reading equipment is not readily available. For up to ten samples, colorimetric detectors are more economical than direct reading instrumentation.

Colorimetric detection of carbon monoxide is based on a reaction with chemicals contained within the detector tube. A color change occurs, and the length of the discoloration (e.g., Drager® white to brownish green) is the measure of concentration.

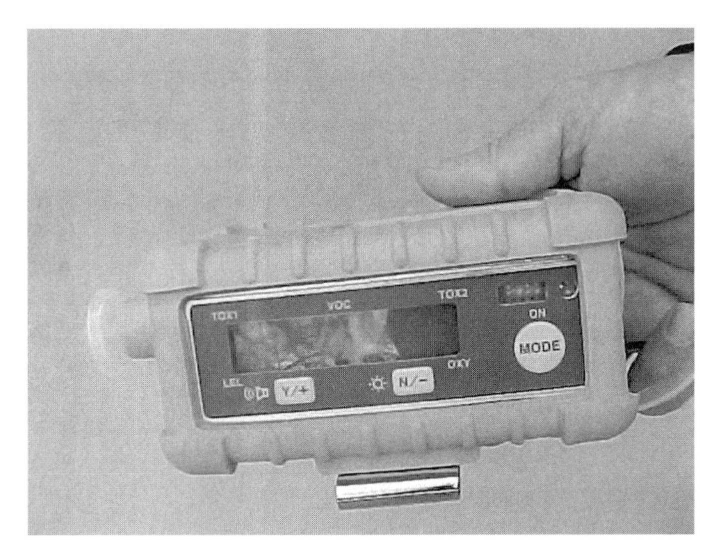

Figure 11.2 MultiRAE® Toxics Monitor–Records and data-logs carbon monoxide along with volatile organic compounds and combustibles.

Sampling requires a hand-held bellows or piston pump and carbon monoxide detector tubes. The pump and tubes must be made by the same manufacturer. The tubes are chemical specific with different ranges (e.g., 2 to 60 ppm and 8 to 150 ppm).

As concentration is difficult to read, results may be off by as much as 50 percent. Although this method lacks accuracy, the range of ± 50 percent has minimal impact.

Although there are several potential interferences and/or cross sensitivities for carbon monoxide detector tubes, they may be filtered with a carbon pre-filter. The potential cross sensitivity chemicals are petroleum hydrocarbons, benzene, halogenated hydrocarbons, and hydrogen sulfide. For example, halogenated hydrocarbons (e.g., trichloroethane) in high concentrations can discolor the indicating layer to a yellowish brown.

The sampling method is summarized as follows:

- Equipment: bellows or piston pump
- Sampling Media: carbon monoxide detector tubes in desired range
- Procedure: insert tube in pump holder with arrow pointing toward the pump, squeeze and release the pump for the number of strokes designed by the manufacturer (1 to 10 times), allowing the pump bellows or piston to fill completely between each stroke
- Analysis: read the concentration on the basis of the perceived length of the color change

HELPFUL HINTS

Take prolific notes, and record sample locations on a building schematic. There is great truth in the old adage "a picture says a thousand words." Record the precise sample locations and sample times, and sketch the sample location wherever possible. Be certain, also, to record the time each data point was collected, conditions, and activities that may impact the interpretation of results.

Be particularly cautious whenever intending to leave data-logging equipment unattended for an extended period of time. Murphy's Law may be instituted during an abbreviated departure from the sample site, particularly where another party may benefit by altered conditions.

INTERPRETATION OF RESULTS

Although the ACGIH recommended limit for carbon monoxide (e.g., 25 ppm) is the simplest approach to interpreting sample results, some investigators choose a more stringent action level. This limit may be two times the outside level in rural environments. It may be 50 percent of the ACGIH limit (e.g., 12 ppm) for 8 hours or a ceiling limit of 25 ppm. The latter numbers are, however, very strict, possibly unrealistic.

Literature, regarding the effects of carbon monoxide, indicates that the lowest level found to effect individuals has been that of smokers exposed to levels of 50 ppm for 6 to 8 hours, and the affect was merely suggested. Tests indicated altered time discrimination, visual vigilance, choice response, visually evoked response, and visual discrimination. Thus, an attempt to reduce an action level below the ACGIH limit is cautious to the extreme.

Where air monitoring is performed with colorimetric detector tubes, only one data point per site will not give an 8-hour average. In this case, the investigator may use the ACGIH 8-hour limit as an indicator that further monitoring is indicated.

SUMMARY

Carbon monoxide monitoring should be performed whenever symptoms are suspect. The source is generally predictable, and sampling should be performed at a minimum at the source at the time when worse-case carbon monoxide levels are anticipated. If the time of exposure occurrence is not clear, long-term data monitoring should be performed. The most commonly accepted action level is 25 ppm averaged over 8 hours. If an 8-hour average is not feasible, one data point of 25 ppm may be used as an indication that further investigation is needed.

REFERENCE

1. Proctor, Nick H., Ph.D. and J. Hughes, M.D. *Chemical Hazards of the Workplace.* J.B. Lippincott Company, Philadelphia. (1978). p. 152.

Chapter 12

FORMALDEHYDE

As one of the twenty-seven molecules identified within the Milky Way and with an estimated production of 1.4 million tons annually in the United States, formaldehyde is everywhere. It is indoors and out, naturally occurring and man-made. It is a by-product of combustion. It is used in the production of home and office products. It is used in cosmetics, and it is in many of our foods both naturally and as a contaminant.

Due to its ubiquitous nature and extensive use, formaldehyde was originally the target health hazard in indoor air quality investigations from the late 1970s until just recently. Known health effects due to low level exposures typically found in indoor air quality include irritation of the eyes, nose, and throat. Symptoms may include watery eyes, burning eyes and nose, and coughing. In more severe cases, there may be lung irritation and bronchospasm.

More elevated acute exposures may include coughing, wheezing, chest pain and tightness, increased heart rate, and bronchitis. There have also been reports of asthma attacks, nausea, vomiting, headaches, and nose bleeds. Exposures in excess of that normally found in indoor air quality situations (e.g., industrial exposures) may result in pulmonary edema, pneumonitis, and death.

Dermal exposures may result in skin irritation, contact dermatitis, and allergic sensitization. In some cases, symptoms involve eczema to the eyelids, face, neck, scrotum, and flexor surfaces of the arms. Sometimes it may involve the fingers, back of the hands, wrists, forearms, and parts of the body exposed to friction.

There are several peculiar health effects of low level formaldehyde exposures encountered in indoor air quality complaints that are not commonly reported in industrial exposures. These include sleeping difficulties, anxiety, fatigue, unusual thirst, dizziness, diarrhea, menstrual cramps, and memory loss. In one instance, a woman thought low level exposures in her mobile home caused her to experience a feeling of persecution. People were allegedly following her, and she attacked an unsuspecting passerby and was dragged away after attempting to strangle the stranger.

According to the ACGIH, formaldehyde is a suspect human carcinogen. A seven-fold cancer risk occurred during industrial operations where exposures were excessive, higher than that of the normal population.[1]

Although a volatile organic compound, formaldehyde is too volatile and low in molecular weight to be captured and analyzed by the methods presented in Chapter 9.

Thus, it must be addressed separately and sampled by different air sampling methodologies.

OCCURRENCE OF FORMALDEHYDE

In 1994, the U.S. EPA reported ambient exposures in urban outdoor air environments to be 11 to 20 ppb. These exposures are in contrast to the over 2.4 million mobile home dwellers that are constantly exposed to an average 400 ppb formaldehyde annually. The range of reported exposures in homes is 0.10 to 3.68 ppm. See Figure 12.1 for formaldehyde exposure levels and associated health effects.

Environmental ambient air exposures are predominantly the result of combustion byproducts with significant contributions from motor vehicle exhaust. Other contributors are power plants, manufacturing facilities, and incinerators. Heavy smog due to urban combustion may reach ambient air levels as high as 0.1 ppm.

Indoors and outside combustion products that contribution to the formaldehyde levels include the following:

- Gasoline
- Wood
- Natural gas
- Kerosene
- Cigarettes

These source contributors are rarely considered significant contributors to the formaldehyde levels indoors.

In office buildings, the most common indoor sources of formaldehyde are off-gassing from building materials and office furnishings. Known building materials that off-gas formaldehyde are formaldehyde-bonded resin products such as foam insulation, plywood, particleboard, pressboard, wall panels, and wood finishes. The most commonly used indoor resin is water soluble urea-formaldehyde. Outdoor resins are phenol- and melamine-formaldehyde.

Other building materials potentially composed of urea-formaldehyde are glass fiber duct board insulation and carpet backings. New office furnishings manufactured with formaldehyde-containing resins (e.g., veneered particleboard desks) are significant contributors to office formaldehyde exposures. Whereas formaldehyde off-gassing of furnishings generally takes place within the first months of purchase, older furniture is rarely a contributor to elevated levels of formaldehyde in the air.

Formaldehyde-containing products may also contribute to office and residential exposures. These may include, but not be limited to, paper products, deodorants, fabric dyes, inks, disinfectants, deodorizers, air fresheners, cleaners, pesticides, preservatives, paints, and permanent press clothing.

In residential environments, many of the same building products and furnishing found in office buildings are contributors to elevated levels of formaldehyde.

Prior to strict manufacturing specifications and state regulations, the highest residential exposures were associated with mobile homes and conventional residences insulated with urea-formaldehyde foam.

Beyond the previously mentioned airborne exposures, individuals inadvertently contribute to their own exposures by applying or using products that contain formaldehyde. Formaldehyde is used in cosmetics, shampoos, pharmaceutical products, and permanent press clothing. If the container does not state the presence of formaldehyde, look for Quaternium. Quaternium 15, a formaldehyde-release agent, and similar preservatives (e.g., diazolidinyl urea) are found in conditioners, deodorant soaps, hairspray, styling mousse, fluoride toothpaste, mouthwash, mascara, talcum powder, hair coloring, and fingernail polish. These should be considered suspect wherein an individual has localized dermal complaints.

The ACGIH 8-hour exposure limit for formaldehyde is 0.3 ppm, and the EPA ambient air action level is 0.1 ppm. The Canadian guideline for formaldehyde exposures in indoor air quality is 0.1 ppm, and the World Health Organization recommends an indoor air quality action level of 0.12 ppm. The National Academy of Science estimates that 10 to 20 percent of the general population is susceptible to the irritating properties of formaldehyde at levels below 0.1 ppm.[2] In combination with other airborne irritants, formaldehyde may also have an additive or synergistic impact of building occupants which could complicate an interpretation of exposures.

Table 12.1 Formaldehyde Exposure Levels and Reported Health Effects

Exposure Level(s)	Health Effects
0.05-1.0 ppm	pungent odor
0.01-2.0 ppm	eye irritation
1.0-3.0 ppm	irritation of eyes, nose, respiratory tract, throat, and upper respiratory tract
4.0-5.0 ppm	unable to tolerate prolonged exposures
10.0-20.0 ppm	severe respiratory symptoms and difficulty breathing
>50 ppm	serious injury to the threshold

Relative humidity, temperature, air movement, and ventilation rates significantly affect formaldehyde off-gassing and airborne exposure levels. For every 10°F increase in temperature, formaldehyde levels double. For an increase in relative humidity from 30 to 70 percent, the exposure levels may increase by 40 percent. Increased air movement will minimize stagnant air pockets and localization of gaseous formaldehyde at the point of off-gassing. Although ventilation flow rates will dilute contaminated air, scenarios where the return air is in the immediate vicinity of the supply air may result in localized recycling and movement of air. The ideal configuration, rarely encountered, is the return air and supply at remote sites to one another.

SAMPLING STRATEGY

At a minimum, samples should be taken within the area of greatest complaints, a non-complaint area, and outside. The investigator should identify these areas and be clear as to the worse-case sites based on the complaints and on odor. As there are no known direct reading instruments capable of measuring formaldehyde at the levels generally found in indoor air quality investigations, instrument screening is out of the question.

For exposure monitoring, samples should be taken within the vicinity of the occupants' breathing zones, not up in the far corner of the ceiling, and all samples should be area samples, not personnel samples. The only occasion when sampling at ground level should be considered is when there is concern regarding exposures to a small child. However, diagnostic sampling is separate issue.

Diagnostic sampling may be performed for source identification. This may be done either during the exposure monitoring or after excessive exposure levels have been determined. In all cases, several samples taken at the same time will assist the investigator to contour an area regarding relative concentrations. Whether taken during or after the initial exposure monitoring, diagnostic samples should also include at least one exposure sample taken at the same time. If the initial sample has already been taken, take another at the same location. This will allow for changed conditions and probable different exposure levels from one sampling period to the next.

At a minimum, samples should be collected during those times anticipated to include excessive exposures, and the air handling system should be operating as it generally is during complaint periods. In an office building, if the complaints occur in the evening when the air handler has been minimized, the samples should be taken in the evening with the air handler minimized. Try to sample during those times and conditions similar to when complaints occur.

At a minimum, samples should be collected for the length of time that will allow detection to the level(s) targeted for concern. If collected by NIOSH Method 2016 at a flow rate of 0.1 liters per minute for 150 minutes, the sample detection will be 0.18 ppm. See flow rates, maximum air sample volumes, and lowest detection limit in Table 12.2.

Other important sampling strategies involve anticipated physical state and interferences. This topic is addressed in the next section.

SAMPLING METHODOLOGIES

Most formaldehyde air sampling methodologies collect on the basis of physical state. Formaldehyde is a highly water soluble gas that readily mixes with water to form formalin, which is a liquid. Formalin vapors may or may not be captured by those methods that are meant only to collect the gas phase, and formalin vapors are

Table 12.2 Formaldehyde Air Monitoring Methodologies

Method	Collection Method	Flow Rate(s) (liters/min.)	Maximum Air Sample Volume	Lowest Detection Limit[A]	Analytical Method(s)	Interferences
NIOSH 2016 EPA TO-11 (gas phase)	/treated silica gel sorbent	0.1-1.5	15 liters	0.18 ppm	HPLC-UV	ozone
NIOSH 2541 (gas phase)	treated XAD-2 sorbent0.	01-0.1	100 liters	0.25 ppm	GC-FID	–
OSHA 52 (gas phase)	treated XAD-2 sorbent	0.01-0.1	24 liters	0.04 ppm	GC-NPD	–
NIOSH 3500 (all phases) solution	2 impingers in line with 10% sodium bisulfite	0.2-1.0	36 liters	0.025 ppm	VAS	phenol and oxidizable organics
NIOSH 5700 (solid phase)	5-micron PVC filter	2.0	1050 liters	0.0003 ppm	HPLC-UV	–
EPA IP-6A (gas phase)	treated silica gel sorbent	0.1	300 liters	0.001 ppm	HPLC-UV	ozone

[A] Based on the published range and maximum air volume.

Analytical Procedures

HPLC-UV	High Pressure Liquid Chromatography/UV Detector
GC-FID	Gas Chromatography/Flame Ionization Detector
GC-NPD	Gas Chromatography/Nitrogen Phosphorous Detector
VAS	Visual Absorption Spectrometry

just as irritating as formaldehye gas. Should formalin become adsorbed onto dust particles, the gas phase collection methods pre-filter dust particles. Where dust may potentially carry formalin that contributes significantly to exposure levels and health effects, choice of a gas collection method will result in incomplete information.

Most of the air sampling methods (i.e., NIOSH 2016, NIOSH 2541, OSHA 52, and EPA IP-6A) will capture formaldehyde in its gaseous phase. They are similar in equipment requirement, collection media, and flow rate restrictions. The differences are in the detection limits (which in indoor air quality are important), sampling duration, laboratory analytical methods, and interferences (or specificity).

Whereas the EPA IP-6A method involves longer sampling periods (up to 50 hours), greater detection limits (0.001 ppm), and may involve ozone interferences, OSHA 52 involves considerably less sampling duration, less detection limits (0.25 ppm), and no interferences. Each of the gaseous phase sampling methods has a slight variation in capability, and in order to conduct a thorough survey, the investigator must consider these differences and weigh them against the sample information sought. A generalized summary of the gaseous phase methods is as follows:

- Sampling Equipment: air sampling pump and tube holder
- Collection Medium: treated solid sorbent (see Table 12.3)
- Flow Rate(s): 0.01 to 0.1 liters/minute (maximum: 1.5 liter/minute)
- Sample Duration: 10 minutes to 15 hours (maximum: 50 hours)
- Detection Limit(s): 0.0003 to 0.18 ppm

Table 12.3 Formaldehyde Solid Sorbent Tubes

Method	Type of Sorbent
NIOSH 2016/EPA TO-11	Silica gel treated with 2,4-dinitrophenylhydrazine (DNPH)
NIOSH 2541 and OSHA 52	XAD-2 treated with 10% (2-hydroxymethyl) piperidine

NIOSH Method 3500 is a good all around method for sampling the gaseous, liquid vapor (e.g., formalin), and solid particle (e.g., formalin adsorbed onto the surface) phases of formaldhyde. Yet, sampling is more difficult. The collection medium is a liquid that is easy to spill (particularly when transferring the liquid into a small plastic container for shipping) and during shipping, poorly sealed containers can leak. It is not unusual for a sample to arrive at a laboratory with only a drop of liquid in the bottom of the container. Furthermore, during extended sampling periods, the collection liquid vaporizes. Although the liquid can be replenished as it vaporizes, the investigator must frequently check the liquid levels during monitoring.

This approach also requires less sampling time while providing a good detection limit. Although present, analytical interferences are slight.

In NIOSH 3500, phenol may be a positive interference. Although low levels of phenol may contribute a 15 percent bias, phenol is rarely encountered in indoor air quality investigations, and separate sampling can be performed to rule phenol out. Slight negative interferences may also result from alcohols, olefins, aromatic hydrocarbons, and cyclohexanone. This method also calls for the use of a 1-micron PTFE membrane filter prior to the two impingers—if the environment is dusty. However, the use of a filter precludes formaldehyde adsorbed into the surface of particles. For this reason, an investigator may choose to use NIOSH 3500 without the filter particularly where there are no excessive levels of dust. NIOSH 3500 method is summarized as follows:

- Sampling Equipment: air sampling pump and two impingers (see Figure 12.1)
- Collection Medium: 1% solution of sodium bisulfite (20 ml. of solution per impinger)
- Flow Rate(s): 0.2 to 1 liter/minute
- Sample Duration: 100 minutes
- Detection Limit(s): 0.025 ppm
- Handling: transfer sampled solution and ship in special 50-ml. low-density polyethylene bottles

NIOSH 5700 was created specifically for sampling wherein formalin-carrying dust

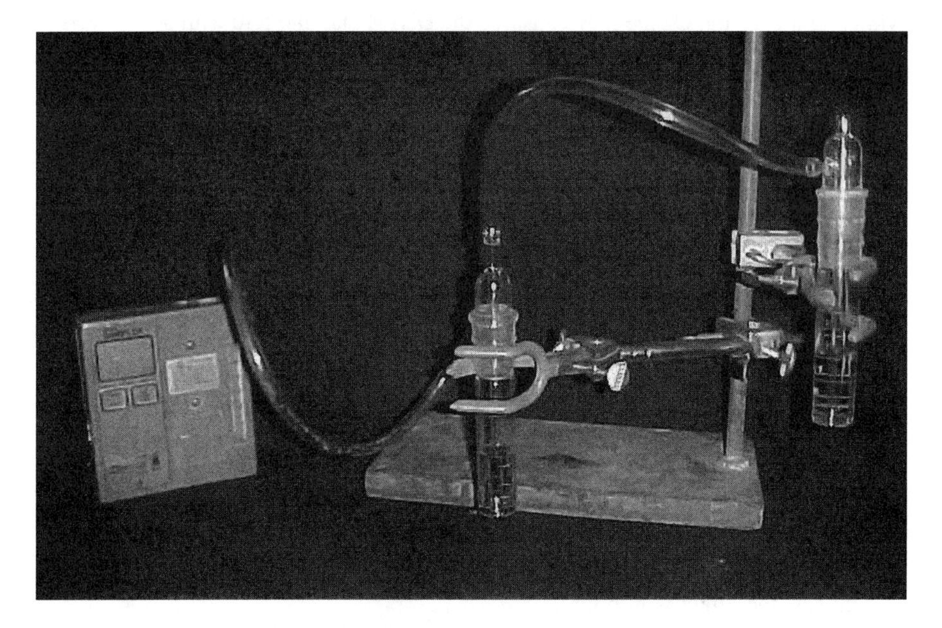

Figure 12.1 Air Sampling Train with Two Impingers

is suspect. In some instances, the investigator may choose to use this method with a gaseous phase method in order to assure collection of all airborne formaldehyde/ formalin. There are no interferences and the detection level is low (e.g., 0.0005 ppm). A summary of this method follows:

- Sampling Equipment: air sampling pump and inhalable dust sampler
- Collection Media: 25-mm PVC filter
- Flow Rate(s): 2 liters/minute
- Sample Duration: 36 to 180 minutes
- Detection Limit(s): 0.025 ppm
- Handling: transfer sampled filter and ship in special 30-ml. low-density polyethylene bottles

See Table 12.3 for a recap of the different methodologies. Sampling equipment is pictured in Figure 12.1.

ANALYTICAL METHODOLOGIES

In NIOSH 2016, EPA TO-11, and EPA IP-6A, where the collection medium is silica gel treated with 2,4-dinitrophenylhydrazine (DNPH), formaldehyde combines with the DHPH on the silica gel to form a 2,4 dinitrophenylhydrazone derivative. The derivative is analyzed by high-pressure liquid chromatography with an ultra-violet light detector. Ozone may consume the DNPH and interfere with the conversion of formaldehyde to the oxazolidine analyte. Thus, where elevated ozone levels are anticipated or known, these methods may not be the best choice for sampling.

Where the collection medium is specially treated XAD-2, formaldehyde combines with the 10% (2-hydroxymethyl) piperidine to form an oxazolidine derivative. The derivative is analyzed by gas chromatography with a flame ionization detector. NIOSH 2541 calls for nitrogen phosphorous detector (NPD) in order to increase the analytical sensitivity, and OSHA 52 specifically directs its use. Although there are no observed interferences with either of these methods, acid mists may inactivate the sorbent, leading to inefficient collection of formaldehyde. Thus, where acid mists are anticipated or known, NIOSH 2541 and OSHA 52 may not be good choices of sampling method.

In NIOSH 3500, also referred to as the "chromotropic acid method," formaldehyde, not a derivative of formaldehyde, is the actual analyte. During sampling, the formaldehyde is settled out by the sodium bisulfite solution. At the laboratory, it is mixed with chromotrophic acid and sulfuric acid. A color develops, and analysis is performed by visible absorption spectrometry. If there is interference by other aldehydes, the effect is minimal.

In NIOSH 5700, the collected dust is extracted with distilled water and mixed with

2,4-dinitrophenyl-hydrazine/acetonitrile (DNPH/ACN). The derivative is then analyzed by high-pressure liquid chromatography with an ultraviolet light detector. This is similar to NIOSH 2016, EPA TO-11, and EPA IP-6A and has the same ozone interference.

HELPFUL HINTS

When performing NIOSH 3500, check frequently for retention of solution and attempt to maintain the level up to 20 milliliters with fresh solution. Upon sample completion, transfer each sampled solution to a separate properly labeled plastic bottle, clean, and add the extra to solution for transport to the laboratory. Seal each bottle with a stretch tape (e.g., electrician's tape) and pull around the seam in the direction that the cap turns to allow for a more air tightfit. Record the level of solution in the container prior to shipping and report it with the air volume for Impinger I (first impinger) and Impinger II (backup).

Given a low flow rate and back-up section or solutions, the investigator may sample for a longer duration, sample greater air volumes, and get better detection limits with reliable results. For instance, if the investigator samples for 5 hours instead of 2½ hours at a flow rate of 0.1 liters per minute, using NIOSH 2016, detection levels can be improved from 0.18 to 0.09 ppm. Most experienced industrial hygienists will not push the envelop. If you do, however, do it with a backup to confirm completeness and reliability of the sample(s).

INTERPRETATION OF RESULTS

There are no federal limits for formaldehyde in non-occupationally exposed occupants of office buildings and residences. For formaldehyde, the OSHA regulatory limits (i.e., 0.75 ppm averaged over 8 hours) for occupationally exposed people greatly exceed exposure levels known to impact human health. Experienced investigators settle on a response action limit whereby remediation or a more in-depth investigation is required.

Exposure action levels may be determined on the basis of health effects, comparison with outdoor environmental levels, or widely accepted limits. The investigator may consider setting an action limit of half that level which is known to cause a given health effect. For example, wherein the primary complaint is irritation of the upper respiratory tract with a known impact level of 1 ppm, the action limit may be set at 0.5 ppm.

Comparative studies should not be overlooked. If the primary complaint is eye irritation, a symptom may be associated with other substances (such as allergens) and/or the published lower limit for health effects are below the normal outdoor ambient levels. For example, formaldehyde is known to cause eye irritation at 0.01 ppm, and the outdoor ambient levels may be 0.1 ppm. In this case, an action level becomes absurd. Thus, the investigator may choose to set an action level on the basis of indoor formaldehyde levels as compared to outdoor levels.

One of the more commonly accepted limits for formaldehyde is 0.1 ppm. This is the EPA outdoor limit and the Canadian suggested limit for indoor air quality. The EPA limit is rarely exceeded outdoors, and the limit of 0.1 ppm encompasses most of the health effects. Some states in the United States have even more restrictive limits. Connecticut has an 8-hour exposure limit of 0.015 ppm, and Vermont has a limit of 0.087 ppm. Many of these more stringent state guidelines may, however, be unattainable.

All things considered, always compare the indoor with the outdoor air samples. A comparison of compliant exposure levels with noncompliant exposures should also be considered. The action level may require adjusting.

SUMMARY

There are pitfalls with each of the sampling techniques. The investigator may use one, or a combination, of these methods, depending on the situation. Consider physical characteristics (e.g., gas verse carried by a solid), detection limits (0.0003 to 0.25 ppm), interferences (e.g., ozone), sampling duration (2 to 50 hours), and ease of sample management. Sampling requirements will vary, depending on each separate circumstance.

An action level for indoor air quality exposures is precarious and relative to other environments. The most commonly accepted limit of 0.1 ppm has boundaries and exceptions as well. With each new case, the investigator plays a new deck of cards!

REFERENCES

1 National Cancer Institute. "Cancer Facts/Risk Factors." http://cancernet.nci.nih.gov/clinpdq/risk/Formaldehyde.html. (22 Dec. 2000).

2 "Formaldehyde Alert!" http://members.tripod.com/gentle-survivalist formaldehyde.html. (22 Dec. 2000).

Chapter 13

PRODUCT EMISSIONS

Indoor nonindustrial exposures to organic compounds are typically two to one hundred times higher than those found outdoors. This is primarily attributed to emissions from construction materials, furnishings, office supplies/equipment, maintenance and cleaning products. Other contributing sources include individual use products (e.g., perfumes and lighters) and outdoor air pollutants.

Indoor industrial exposures to volatile organic compounds are generally ten to one hundred times those found in nonindustrial environments. This extreme difference raises a jaundiced eye whenever an environmental professional responds to office building complaints. Exposures are so much lower in nonindustrial environments. Yet, complaints are more prominent.

The problems/complaints in office buildings appear disproportionate to the lack of problems/complaints in industry. Some feel that the discrepancy lies in the medley of chemicals. Office environments generally consist of up to three hundred chemicals, an amalgam that far exceeds most industrial exposures that might consist of up to ten components. So, due to complexity of building construction/furnishing emissions, architects, construction managers, building owners, state agencies, and other organizations have begun to seek supplies with minimal product emissions.

Space management plans have taken into consideration the building design, number of occupants, and location of occupants. Yet, as less space becomes available, crowding culminates in more congestion, greater product emissions.

Renovation activities, pesticide treatments, maintenance, cleaning products, and outdoor repairs/roofing activities contribute to the quality of the indoor air. Some construction materials and many furnishings concentrate or amplify airborne chemical contaminants that are introduced by other materials or treatments due to cleaning products, pesticide applications, and maintenance activities. Whereas they may not be a direct source of exposure, these materials may contribute indirectly.

Construction practices and building commissioning/recommissioning are also included in the plethora of issues that impact the burden of airborne chemicals. New building construction and recently renovated indoor environments have been found to have volatile organic chemical levels similar to those found in industry (e.g., up to 30 mg/m^3 of volatile organic compounds). Surprisingly, the higher levels are associated with residential structures more than those found in office buildings.

On an increasing frequency, products are being selectively manufactured to minimize or eliminate emissions all together in nonindustrial indoor environments. Market awareness is the driving force. Product emissions testing is the first line of defense!

USES FOR PRODUCT EMISSIONS TESTING

Product emissions testing is used by manufacturers for product information and by environmental professionals for predicting product emission contributions to indoor air environments. Although the ultimate concern is product emissions (frequently referred to as off-gassing) in new or renovated office environments after installation, a more practical approach is required testing prior to any significant purchases.

Testing is performed by some manufacturers for quality control. The manufacturer may have testing performed at various stages in the production process to determine the 24-hour emission factor for a predetermined time period. This is generally performed, if not intermittently, upon the finished product. With predetermined limits, the manufacturer addresses and corrects manufacturing discrepancies to minimize the sale of products that may contribute excessive irritant/hazardous substance levels to the indoor air environment. The manufacturer, also, uses this information to identify process variances which may contribute to elevated emissions from their products. In this manner, manufacturing processes are altered to accommodate the need for reduced off-gassing. Products which frequently receive considerable attention are particleboard (e.g., formaldehyde) and carpeting (e.g., 4-phenylcyclohexene). Some other products implicated in product emissions are listed in Table 13.1.

Product testing is performed to meet special labeling restrictions. The Carpet and Rug Industry has a pass/fail indoor air quality labeling program for its participants. Upon successfully passing the emissions test, the participant is approved to

Table 13.1 Components Which Result in Product Emissions[1]

VOLATILE ORGANIC COMPOUND EMISSIONS

- Paints
- Fabrics and fabric treatments
- Cushions (polyurethane/polystyrene foam and polyester stuffing)
- Plastics (various plasticizers)
- Adhesives
- Cleaning solvents

FORMALDEHYDE EMISSIONS

Particleboard	Glues
Plywood	Resins
Pressboard	Insulation
Paneling	Foam
Carpeting/carpet backings	Laminates
Dyes	
Plastics/moldings	
Household cleaners	
Stiffeners	
Wrinkle-resistant fabrics	
Water repellents	

place a green and white label on their carpeting that states "Carpet and Rug Industry Indoor Air Quality Testing Program."

Carpet Component: 4-phenylcyclohexene
(binder, backing, and glue)

Copiers and laser printers also have a labeling program. It is referred to as the "Blue Angel." Only products meeting strict emissions limits may be labeled. Although the labeling originated in Germany, there is a growing interest internationally.

Some environmental professionals address product emissions prior to construction and purchase of furnishings. The state of Washington has instituted an East Campus Plus Indoor Air Quality Program of all state-owned buildings. It involves construction practices, selection of materials, methods for commissioning renovated/newly-constructed buildings, and acceptable office building limits for total volatile organic compounds and formaldehyde. The Program requires all major construction materials and office furnishings to undergo environmental chamber testing and, based on predicted levels, ensure the products will not, in combination, exceed the acceptable limits. Any products exceeding the limits require pre-installation airing. The means used for predicting the total burden is through an EPA computerized algorithm with emission rates and other building-specific features (e.g., volume, air changes, etc.).

Toxicological testing that is based on environmental chamber technology and animal inhalation studies is performed to address sensory irritation caused by airborne chemicals. Product emissions are introduced into a chamber housing laboratory animals. The breathing patterns and other clinical responses are then monitored. Challenges are singular or multiple.

There have been isolated instances of human irritation testing using environmental chambers. For example, a researcher who has received considerable attention internationally used chamber challenge testing to develop a dose response relationship for discomfort due to volatile organic compounds. See Table 13.2 for a summary of his findings.

On this order, there have been attempts by some researchers to perform challenge testing without the benefit of a chamber and/or known, monitored levels of exposure. As the actual chemical exposure levels are poorly controlled, the results become questionable. Furthermore, there is a concern for life-threatening patient reactions, and a medical emergency must be considered when performing the tests. For these reasons, human challenge testing is rare in the United States.

MEASURING UNITS AND EXAMPLES

Product emissions and their impact on a confined office/building environment are measured by one of several approaches, each with an emphasis on and targeting different information. A few of the more commonly used units include emission factors, emission rates, and predicted air concentrations.

Table 13.2 Summary of Research Findings on Effects of Total Volatile Organic Compound (TVOC) Mixtures[2]

TVOC (mg/m^3)	Health Effects/Irritancy Response
<0.20	No response
0.20 - 3.0	Irritation and discomfort
3.0 - 25	Discomfort (probable headache)
>25	Neurotoxic/health effects

An "emission factor" is the amount of chemical that is emitted from a product at a specified time in the life of the product.[3] For example, a product may be tested immediately after production, just prior to shipment, or prior to installation. The time is defined, and numbers vary during the manufacturing evolution and life of a product. Solid product emissions are based on exposed surface area(s), micrograms per square meters-hour ($\mu g/m^2$ -hour). Liquid products may be based on relative mass, micrograms per gram-hour (μg/g-hour), and whole units may be reported in composite form, micrograms per composite-hour (μg/composite unit-hour). In the latter, the composite may be a single workstation or a group of furnishings. An example follows:[1]

System office furniture

- Formaldehyde: 802 to 3,780 mg/workstation-hour
- Volatile organic compounds: 160 to 45,000 mg/workstation-hour

Chairs

- Formaldehyde: no detection to 1,670 mg/chair-hour
- Volatile organic compounds: 159 to 450 mg/chair-hour

Tackable acoustical partitions (with phenol-formaldehyde treated fiberglass insulation)

- Formaldehyde: 0.158 to 0.37 mg/m^2-hour
- Volatile organic compounds: 0.006 to 0.074mg/m^2-hour

See Table 13.3 for additional findings. Each of these studies is limited in scope therefore not to be considered conclusive and/or all encompassing regarding all products. Variations will occur between product types, manufacturers, formulations, and production processes.

Predicted air concentrations are location specific. Calculations are based on the product's emission rate as well as specific building environmental components.[3] These components include air movement, amount of make-up air, air volume capacity of the room(s), and level of occupant activity. The complexity of the algorithm lends itself to computer modeling. The EPA Exposure Model is widely used, and its predictions have been validated.

Table 13.3 Emissions from Furnishings/Related Products[1,4]

Product	Emission Rate (μg/m²-hour)
TOTAL VOLATILE ORGANIC COMPOUNDS	
Solvent-based adhesives	up to 17,000,000
Water-based adhesives	up to 2,100,000
Furniture spray polish	300,000
Wood stain	17,000
Polyurethane lacquer	6,000
Plywood	up to 2,400
Polystyrene foam	up to 1,400
Particleboard/fiberboard	up to 150
Hardboard	30
Medium density fiberboard	40
FORMALDEHYDE	
Wood products	170 to 900
Insulation	16 to 26
Wall coverings	20 to 600
Textiles	not detected to 3,000

SAMPLING METHODOLOGIES

There are no hard, fast rules for sampling. Indoor sources of emissions vary widely in both the strength of their emissions and in the type/number of compounds emitted. Differences in the emission rates vary to several orders of magnitude even with the same type of material. Therefore, the amount of material required for analysis will vary by the experience of the commercial laboratory and their directives on a case-by-case situation.

Then, too, entire modules may require special handling practices, and a large test chamber that limits the choice of laboratories that are capable of performing the analytical process. The environmental professional should be wary of the time the sample is extracted or packaged, precautions for containment, and transport environment when planning to ship a sample for analysis. All this should be arranged and coordinated through the laboratory of choice prior to sample collection.

ANALYTICAL METHODOLOGIES[5]

Although there is no standard protocol for environmental chamber emissions testing, the American Society for Testing and Materials (ASTM) published *Standard Guide for Small-Scale Environmental Chamber Determinations of Organic Emissions from Indoor Materials/Products* in November 1990. They clearly state that this is a guide only and differences in approach will occur from one researcher/laboratory

Table 13.4 Formaldehyde Emission Factors from Finished Wood Products[1]

Material	Concentrations (μg/m^3)
Medium density fiberboard	970
Unfinished particleboard	up to 809
Finished particleboard	up to 719
Finished medium density fiberboard	up to 246
Water-damaged chipboard	48
Hardboard	up to 30
Medium density fiberboard	up to 14
Non-water-damaged chipboard	10

to the next. A distinction is made between small and large chambers. Small chambers (smaller than 5 cubic meters in volume) are not to be used to evaluate applications (e.g., spray painting activities).

Products (e.g., small samples or whole furnishing modules) are placed in or material (e.g., paint) applied inside a contained, controlled test chamber. The humidity, temperature, chamber air exchange rates, and air movement are controlled/constant. Air monitoring is performed at various predetermined intervals at the headspace (e.g., air immediately above the sample) of either closed containers or dynamic flow-through headspace. Collection is performed for a minimum of the following:

- Volatile organics
- Semivolatile organics
- Polar compounds
- Nonpolar compounds

Monitoring methods may involve capture using a thermal desorption tube or other media as required for the material of interest. Analyses are typically performed by GC/MS.

INTERPRETATION OF RESULTS

In the absence of federal standards for product emission limits, there are a few industry guidelines, researcher recommended guidelines, and occasional state/private building owner standards.

As indoor air quality concerns increase and the public expresses greater interest in product emissions, more industries are expected to set their own criteria for products with low emission rates. The carpet industry was one of the first to show an interest. In order to qualify for the Green Label, carpets must not exceed specified emission factors as measured over a 24-hour exposure period. These emission factor limits are:[6]

- Volatile organic compounds: 0.5 mg/m^2-hour
- Styrene: 0.4 mg/m^2-hour
- Formaldehyde: 0.1 mg/m^2-hour
- 4-Phenylcyclohexene: 0.05 mg/m^2-hour

In Germany, environmental labeling focuses on the environmental acceptability of indoor products, based on chemical emissions and other environmental concerns (e.g., waste generation and recycling). A "Blue Angel" emissions criteria is provided for copiers and laser printers. On the international market, manufacturers are perceiving the growing significance of emissions testing and seeking Germany's "Blue Angel" label for their copiers and laser printers. The maximum allowable emissions are based on environmental contributions and must not exceed the following:[7]

- Dust: 0.25 mg/m^3
- Ozone: 0.04 mg/m^3 (0.02 ppm)
- Styrene: 0.11 mg/m^3 (0.025 ppm)

A researcher with the Environmental Protection Agency suggested certain product default values wherever predicted air modeling cannot or is not going to be performed. These values are for volatile organic compounds only and are based on emission rates:[8]

- Flooring materials (including carpeting): 0.6 mg/m^2-hour
- Wall materials: 0.4 mg/m^2-hour
- Moveable partitions: 0.4 mg/m^2-hour
- Office furniture: 2.5 mg/workstation-hour
- Office machines (central): 0.25 mg/hour-m^3 of space
- Office machines (personal office): 2.5 mg/hour-m^3 of space

The Washington "East Campus Plus Indoor Air Quality Program" has a set of permissible limits for airborne composite sampling. These limits are as follows:[8]

- Volatile organic compounds: 0.5 mg/m^3
- Total particles: 0.05 mg/m^3
- Formaldehyde: 0.06 mg/m^3 (0.05 ppm)
- 4-Phenylcyclohexene (carpet only): 0.0065 mg/m^3

The state of Alaska has adopted these criteria for new construction projects, and California Proposition 65 requires products to be labeled if emissions exceed defined limits. At the present, most of the state limits are principally for new construction on state-owned buildings. Privately owned buildings are internally managed with little or no guidelines.

REFERENCES

1. Franke, Deborah, et. al. Furnishings and the Indoor Environment. [Draft to be published in the *Journal of the Textile Institute*] Air Quality Sciences, Inc., Atlanta, Georgia, 1995.
2. Molhave, L. "Irritancy of Volatile Organic Compounds in Indoor Air Quality." Paper presented at the Fifth International Conference on Indoor Air Quality and Climate in Toronto, Canada, 1990.
3. Air Quality Sciences. "Defining Product Emission Measurements." [Bulletin] Air Quality Sciences, Inc., Atlanta, Georgia, 1995.
4. Black, Marilyn, Ph.D. "Volatile Organic Compounds in the Indoor Environment." [Bulletin] *Indoor Environment '95*, May 1-3, 1995.
5. American Standard and Testing Materials (ASTM). *Standard Guide for Small-Scale Environmental Chamber Determinations of Organic Emissions from Indoor Materials/Products.* ASTM D5116-90. 1990.
6. Air Quality Sciences. How Do CRI's Carpet Emissions Criteria Compared to the State of Washington's Purchase Specifications? [Newsletter] *AirfAQS*, Air Quality Sciences, Inc., Atlanta, Georgia. Fall 2(1):4 (1994).
7. Air Quality Sciences. Office Machine Manufacturers Seek Germany's Blue Angel Certification. [Newsletter] *AirfAQS,* Air Quality Sciences, Inc., Atlanta, Georgia. Winter 2(2):1–2 (1995).
8. Air Quality Sciences. What Guidelines Exist for Chemicals and Particles? [Newsletter] *AirfAQS*, Air Quality Sciences, Inc., Atlanta, Georgia. Summer 1(4):2 (1994).

Section IV

DUST

Chapter 14

FORENSICS OF DUST

Since as early as the late 1800s, scientists have used forensic microscopy in crime detection. Pollen typing has been used to determine the source areas for illegal shipments of marijuana. Crime scene soil samples have been used to locate the source of the material. Clothing fibers are traced by fiber type and special dyes. Hair can be differentiated as to species (e.g., human, dog, or cat) and distinct color, texture, and thickness. Dust found at a crime scene sometimes contains evidence as to an association with certain industrial activities.

Only recently has forensic microscopy been recognized as a tool in indoor air quality investigations. Without forensic microscopy, identification of unknowns was limited. The investigator would develop a theory as to the dust component that caused health problems and test the theory. Not only was this time consuming and expensive, but the actual causative agent was often overlooked.

If building occupants complained of allergy symptoms, an investigator automatically assumed the problem was molds. Even when sample results did not support the theory, the investigator may persist and state that the sampling methods are faulty. This scenario often culminates in an extensive search for the ubiquitous mold and, in many cases, destruction of walls and flooring in the frantic search for the hidden demon. If one looks hard enough, behind enough walls and enclosures, an investigator will eventually locate molds.

With forensic microscopy, the allergenic dust in an occupied space can be characterized. Not only can pollen, mold spores, algae, and insect parts be identified, but an experienced microscopist can characterize rodent, bat, cat, and dog hairs as well. The microscopist can also quantify the population density as normal or excessive. For instance, the microscopist may identify excessive amounts of rodent hairs. When pressed for more information, the microscopist may come back with, "More than normally observed in occupied spaces, typical of rodent-infested barns."

Forensic microscopy has also been used to identify other components of dust that may cause non-allergy health problems. For instance, chemicals adsorbed onto the surface of particles may be identified (e.g., formaldehyde on dust particles). Pharmaceutical dust can be identified from previous manufacturing facilities (e.g., amphetamines). Fibers that may cause lung irritation (e.g., treated glass fibers) and/or long-term health effects (e.g., asbestos) can be identified. Toxic minerals (e.g., silica) and paint components (e.g., lead chromate and fungicides) can be identified.

Suspect materials may be either confirmed or denied. For instance, in one case, white spots on surfaces implicated paint as the source of indoor air quality, yet the spots were silicon.

Forensic microscopy can be used as a tool in identifying particles and chemicals. The list goes on!

OCCURRENCES OF FORENSIC DUST

In 1972, The McCrone Institute performed a study to determine settling rates of dust on surfaces. They found that nearly 1,000 particles per one square centimeter settled hourly. The particles were all in excess of 5 microns in size. The calculated settling rate for dust was thus found to be 24,000 particles per square centimeter per day. Typical dust in indoor air quality was also found to consist of human epidermal cells, plant pollen, human/animal hairs, textile fibers, paper fibers, minerals (from outdoor soils and dust brought indoors), and a host of other materials that may typify a given environment (e.g., fly ash from the gas-burning furnace in the building).[1]

A study by Cornell University suggested that indoor air quality problems were caused by glass fibers.[2] Possible sources of airborne glass fiber exposures include, but are not limited to, fireproofing in air plenums, ceiling tiles, duct board, and furnace filter material. For a few photographic examples of different source glass fibers, see Figure 14.1.

One example of glass fibers in indoor air quality involves a residential occupant. A woman complained of a home-related itch. Her doctor speculated the probable cause was glass fibers. Subsequently, settled dust samples were collected from various areas around the house, and bulk samples were taken of various building/furnishing materials known to have fibers. Each of the settled dust samples was found to contain large amounts of glass fibers that were impregnated and covered with globules of a pink resin. None of the bulk glass fiber samples matched. The investigator returned for additional samples and tracked down insulation (e.g., batting) in the enclosed wall spaces that had the same appearance as the settled fibers in the dust. Thus, the insulation was the confirmed culprit. Upon further investigation, the means by which the enclosed insulation entered into the occupied space was determined. The air movement caused by leaking air ducts disturbed the surface of the interior insulation and picked up the fibers, distributing them throughout the residence through the air supply vents.[3]

Some people say they are allergic to dust, but dust varies in composition. There is a considerable difference between barnyard dust verses dust in a conditioned building. It is the composition that causes health problems, not the dust itself. For a generalized listing of dust components and some representative photomicrographs, see Table 14.1 and Figure 14.2.

It should also be noted that dust composition is in a constant state of flux, even in conditioned office spaces where there are no internal sources of dust (e.g., molds growing, cockroaches, and rodents). People bring in dust on their clothing, shoes, For example, an investigator was attempting to recreate dust exposures that occurred

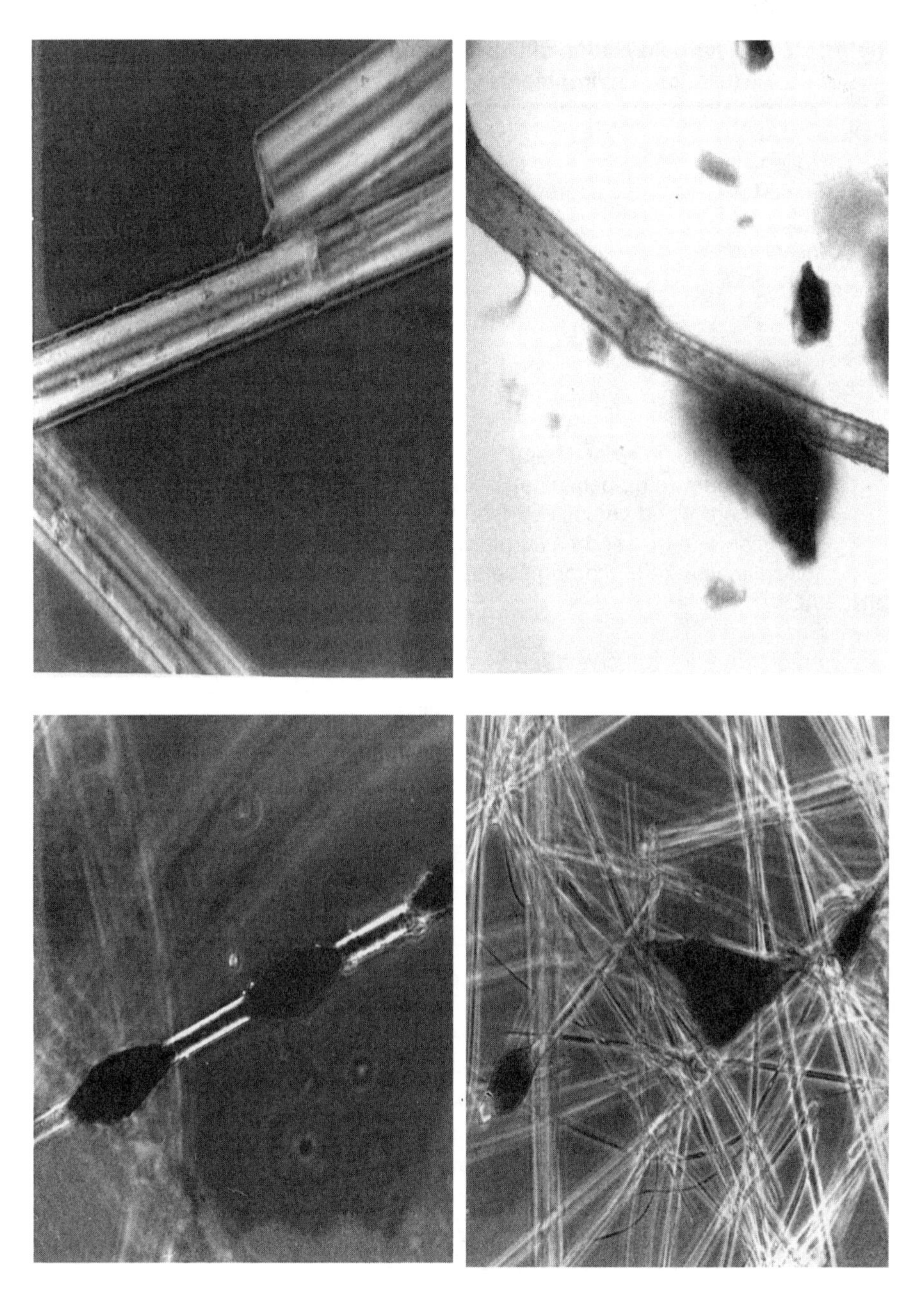

Figure 14.1 Photomicrographs of glass fibers from different sources, magnified 400x. They include: (top left) untreated fiberglass, (top right) duct board, (bottom left) duct board with coating material treated using xylene and sulfuric acid to affect a color change that tags free aldehydes, and (bottom right) thermal insulation with asphalt-impregnated binder.

Table 14.1 Characterization of Dust Components In Indoor Environments

BIOLOGICAL
Pollen
Fungal and bacterial spores
Algae
Insect parts
Skin cells
FIBERS
Hair (e.g., human, cat, or dog hair)
Clothes fibers
Paper fibers
Spun fibers (e.g., glass fibers)
Mineral fibers (e.g., asbestos)
Wood (hard wood versus soft wood)
Plant fibers (e.g., seed hairs, blast/leaf/grass fibers)
Miscellaneous (e.g., carbon fibers, feathers, spider webs, etc.)
MINERALS
Soils
Amorphous versus crystalline
OTHERS
Soot and ash
Metal fumes
Paint
Explosives
Pharmaceuticals
Drugs

Excerpted from *Forensic Microscopy*.[4]

and body surfaces. They bring components of dust from the environments they live, shop, and play in. Yet, it is unreasonable not to anticipate internal sources as well. With consideration for all possible sources, one dust sample may have several allergens and other components that may cause non-allergen health problems. Thus, an investigator should be open to all possibilities.

SAMPLING METHODOLOGIES[5,6]

Although there are a few published approaches, most procedures should be worked out between the analytical laboratory and the indoor air quality investigator. Yet, keep in mind, methods appear simple, but if not completely thought out, these simple methodologies can be misconstrued or misinterpretation. Plan a strategy, and stick to it. If disallowed access, don't take a sample just to take a sample.

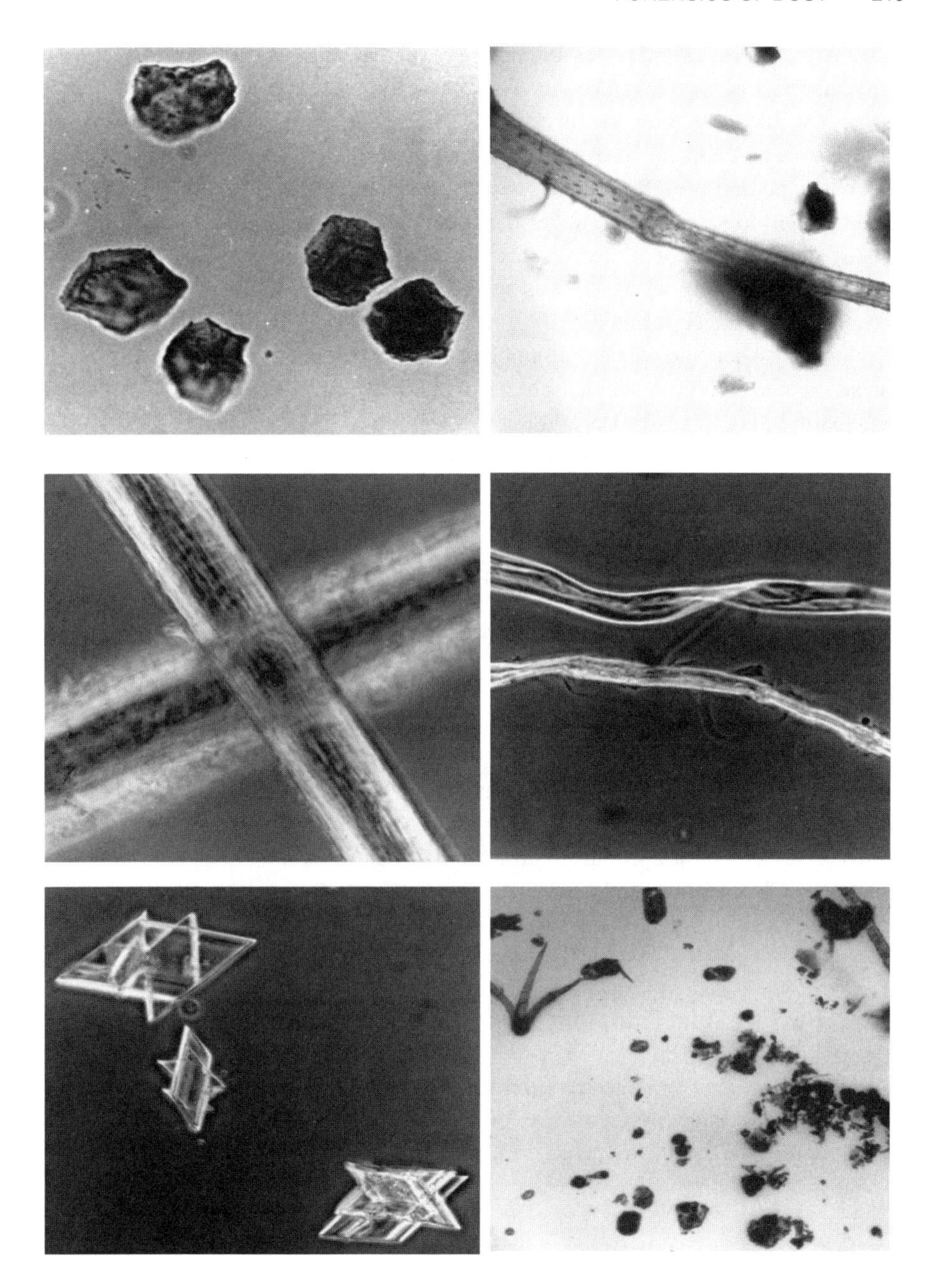

Figure 14.2 Photomicrographs of identified dust component, magnified 400x. They are epithelial cells (top left), an insect leg (top right), hair fibers (middle left), clothing fibers (middle right), crystalline mineral formations (bottom left), and a general overview of environemental dust (bottom right). The latter shows minerals, spores, wood fibers, pollen, and plant hair.

For example, an investigator was attempting to recreate dust exposures that occurred two years prior to sample collection. This was a litigious case whereby the defense attorneys would only allow the investigator to take carpet and/or dust samples from under old filing cabinets. The defense attorneys refused to permit dust collection from above ceiling tiles. So, the investigator took samples from locations where he was permitted to sample (e.g., under the filing cabinet). The forensic samples predictably disclosed minimal dust. In this situation, the investigator wasted time and money sampling only where he was permitted access, not where good judgment dictated the sample be taken.

Consider when and where the sample should be taken. For historic surface dust that has settled over an extended period of time, the investigator should consider areas that frequently get overlooked during cleaning (e.g., ledges above doors, window frames, picture frames, above ceiling tiles, and air supply/return vents). Carpets and upholstery generally retain dust over the passage of years, and if a picture of the past is required, such as in litigation, dust from under the carpet (e.g., bulk sample) or within the upholstery (e.g., micro-vacuum sample) should be collected.

For a more recent settled dust sample, the investigator may want to determine the last time an area was cleaned, record the date/time, and take a sample in an area that has been cleaned. If an area has not been confirmed as having been cleaned, assume it has not. Areas that typically get missed are tops of computers, bookshelves, lamps, and memorabilia.

For airborne dust, the investigator has several means for taking air samples. Air samples would represent dust components and levels that the occupant was breathing during the sampling period.

Methods herein are provided for settled surface dust sampling, airborne dust sampling, bulk sampling, and textile/carpet sampling. Choose the method which is most appropriate to a given situation.

Settled Surface Dust Sampling

Settled dust may be collected from smooth surfaces (e.g., desk tops) and rough surfaces (e.g., carpeting) by any number of techniques. Some require specialty supplies. Others require the use of that which is readily available at a local retail store.

Specialty Tape

Specialty tape may be purchased from microscope supply venders. The tacky material is minimal and does not hold the collected material such that it becomes difficult or impossible to remove. The taped material is retained by affixing the tacky surface to a clean surface and placing it into a fiber-free envelope/plastic bag for transport to the laboratory. At the laboratory, the microscopist will "pluck" the material from the surface of the tape by using a special micromanipulation device (e.g., fine tungsten needles with a tip measuring 1 to 10 microns in diameter). With this technique, the dust components may be isolated and identified individually.

Clear Tape

A clear tape (e.g., 3M Crystal Clear®) is available in some office supply stores. This tape is not the usual tape that one can see through when it is affixed to paper. It is a clear tape where the writing on the carrier is readable. The tape is touched to a surface, and dust particles adhere to the sticky portion of the tape. This is then placed immediately onto a microscope slide with or without a stain.

The drawback to this method is that once affixed to the slide, the collected sample cannot be further manipulated and stained without great difficulty (e.g., treating the surface of the tape with a solvent). Then, too, if there is excessive material on the tape to reasonably distinguish individual particles, the sample may again require special processing. With these limitations in mind, the environmental professional may choose to use this technique only for screening and gross examination of material or for confirming the presence of a suspect material for which the microscope slide has already treated with the appropriate stain.

However, if the investigator should choose to use the clear tape, there are means available to manage the otherwise irretrievable sample. The particles that have adhered to the tape may be removed by lifting the tape, applying a small drop of benzene, and using a fine needle to make a small ball of the adhesive trapped particle(s).[6] The trapped material can be withdrawn and the ball of adhesive removed chemically. This process is tedious and choice of a more easily manipulated collection media is desired whenever feasible.

Post-it Paper [7]

Post-it paper is excellent for sample collection as it is easily obtained and inexpensive. The sticky surface of a Post-it is pressed onto the settled dust, the paper folded into itself (with the sticky portion inside), and shipped to the laboratory of choice in a plastic zip-lock bag. Analysis may be performed by particle picking material from the sticky surface or by scanning electron microscope while the particles are still on the paper.

Micro-vacuuming[8]

Micro-vacuuming has been receiving a considerable amount of attention, particularly for asbestos-contaminated settled dust. A vacuum pump is used to collect dust particulate within a 100-square centimeter surface at a recommended flow rate of 2.0 liters per minute. This method is particularly useful in dust collection from irregular surfaces (e.g., carpeting).

The detection limit of this methodology is 150,000 structures per square foot [or 161 structures per square centimeter (structures/sq. cm.)] as determined by transmission electron microscopy (TEM). Concentrations over 1,000 structures/sq. cm. are considered elevated, while levels over 100,000 were used to indicate an abatement project barrier has been breached. There are no regulations that provide acceptable/non-acceptable limits.

Airborne Dust Sampling

Airborne dust capture may be preferred over settled dust collection in order to determine the existing suspended particulates (not the existing and previously deposited) and to collect some of the smaller material which may not have settled out from or become resuspended in the air due to the size and/or shape of the material of possible concern.

Air-O-Cell Cassette

An Air-O-Cell cassette is a special particulate sampling device. It consists of a treated microscope slide that is contained within a cassette. The slide is treated with a sticky substance onto which mold spores, pollen, insect parts-and-pieces, skin fragments, hairs, and fibers will adhere upon impaction onto its surface. Its most common use has been sampling for nonviable mold spores, mold fragments, and pollen.

The end seals on the cassette are removed, and the cassette is connected by flexible tubing and a ½-inch to ¼-inch converter to an air sampling pump. The pump flow rate is calibrated. Air is sampled for an abbreviated period of time, cassettes are resealed, and samples are sent to a laboratory for analysis.

A summary approach is as follows:

- Equipment: air sampling pump
- Collection Medium: Air-O-Cell cassette
- Flow Rate: 15 liters/minute
- Recommended Sample Duration: 1 to 10 minutes, based on anticipated loading

Anticipated loading is based on environmental conditions and anticipated loading. Where excessive loading occurs on the slide, enumeration becomes difficult if not impossible. In the latter case, samples may be significantly underestimated and difficult to identify.

In clean office environments and outside where there is very little dust anticipated, sampling should be performed for 10 minutes. In dusty areas and/or areas where there is considerable renovation, a 1 minute sample should be considered. Indoor air environments where there is moderate dust or where considerable levels of mold spores (e.g., greater than 500 spores) are anticipated, the sampling duration should be reduced accordingly (e.g., 6 to 8 minutes). Experience will be the investigators best guide.

Membrane Filters

Air sampling may be performed for suspended dust by using a membrane filter that is contained within a cassette and an air sampling pump. This is a dry sampling method and will desiccate, damage the more fragile components in dust.

A summary approach is as follows:

- Equipment: air sampling pump
- Collection Media: cassette containing a filter
- Flow Rate: 1 to 15 liters/minute
- Recommended Sample Duration: 100 minutes (i.e., fragile biologicals), 135 minutes (i.e., high flow rate collection of nonfragile material) to 2,000 minutes (i.e., low flow rate of nonfragile material)
- Recommended Air Volume: 100 to 2,000 liters

Recommended filter types include, but are not limited to, the following:

Polycarbonate filter—Using a stereomicroscope, the microscopist may selectively isolate and pluck material from the surface of the filter.

Fiberglass filter—The microscopist may slice the filter and look at the material through an unspecialized microscope (e.g., not a phase contrast microscope) that will allow a view of the material on the surface of the fiberglass while the light passes through the thinned-out fibrous backing.

Mixed-cellulose ester filter—The microscopist may melt the filter (in a procedure similar to that of asbestos air sample filter analysis) using vaporized acetone. This method is not recommended in most cases where desiccation and/or destruction of the sought after material may occur. If the material is an unknown, this approach will disallow identification of content.

Keep in mind that each of the above filters has different pore sizes, depending upon the manufacturer and the various specifications. One filter type may come with several choices as to pore size and/or particle retention capabilities. The smaller the pore size, the more expensive the filter. Be certain to obtain one that will capture particulates down to 1 micron in diameter or better. All of the above-mentioned filters can be purchased with a minimum of 1 micron and down to better than 0.025 microns.[8] Unless electron microscopy is to be performed, the latter is unnecessary.

The air sampling flow rate should be adjusted to produce as large a sample volume as possible within the time period desired while keeping the upper limit within a flow which will not cause damage to the filter or desiccate biologicals (e.g., mold spores and bacteria). In most cases, a flow rate of 1 liter per minute may be used for biologicals, and 15 liters per minute should not destroy the filter.

Larger air volumes provide more representative samples. Although total sampled air volumes have been as low as 100 liters, a collection of 2,000 liters may prove more valuable. Then, too, the biologicals stand a greater chance of being damaged with longer sample durations. So, the environmental professional may choose to take a minimum of two samples per site where biologicals are a possibility. The two samples may represent the lower range and the larger volume of sample. Based on the method of analysis, in most cases, the air volume cannot be excessive. Like a dump truck of soil sample, the limitation of the air sample volume is relevant only in the provision of a manageable sample size.

Even if the microscopist is not intending to quantify the results in particle per cubic meter, air volumes should be recorded in the off-chance that the volume may be used later (e.g., where asbestos fibers are identified, the airborne fiber counts may be determined). The air volume, however, will rarely be relevant to the microscopist.

Cascade Impactors

An adhesive film may be placed on the surface of each stage of an impactor, and the sample collection time may be limited while separating the collected material by size. As the particles tend to impact singularly, the microscopist may analyze each adhesive film directly. Separation and isolation are already completed by this sampling method.

A summary approach is as follows:

- Equipment: air sampling pump
- Collection Media: cascade impactor with special adhesive film
- Flow Rate: 28.4 liters/minute
- Recommended Sample Duration: 5 to 10 minutes

Other Methods

Other air sampling techniques that have been used include impingers and cyclones. Processing these samples may be more involved (e.g., time consuming and expensive), yet impinger and cyclone samples allow for dilution and separation of the collected material. They are, however, limited in their ability to collect only certain particle sizes. The impinger collects particles greater than 1 micron in diameter, those that are visible under the light microscope. On the other hand, the cyclone collects nonfibrous particles of less than 5 microns. Anything greater than 5 microns and/or fibrous in nature is likely to be overlooked.

Bulk Sampling

An alternative to the specialty tape and problems associated with the clear tape is "bulk dust" sampling. When dealing with large deposits of dust (or dirt), bulk sampling becomes the most feasible approach. Define an area where the collection is indicated and scoop/scrape the dust from the surface, using a fiber-free (e.g., a cellophane envelope), contaminant-free scraper (e.g., a stainless-steel spatula). At the laboratory, large samples are homogenized, and a representative sampling of the entire mix will be extracted. For this reason, when deciding how much to collect, the environmental professional may wish to restrict sample sites and limit collection to clearly defined, distinct areas. For instance, a specific air supply louver in a complaint area and settled dust from a recently cleaned table surface are clearly defined, delineated sample locations.

Then, too, a building material or structural component may appear to be contaminated with an unidentified substance that has become part of its substrate. For instance, a weakened spot on a steel beam may be associated with an unidentified material, or gypsum board may have what appears to be a microbial growth. If possible, collect a piece (e.g., at least a 4-inch square surface) of the substrate. Place it in a plastic baggy and ship with instructions to the lab that you wish to know what the associated material is and describe its physical appearance as best you can. If the instructions are incomplete, the laboratory might just miss the point of the request. Be clear and concise in your instructions!

Textile/Carpet Sampling[9]

Microvacuuming of textiles and carpeting has proven inferior to direct extraction and sonication of textiles and carpeting. A comparative study was performed of asbestos contamination of carpets that demonstrated the difference between the two methods. See Table 14.2.

Textile/carpet sampling involves cutting a piece from the textile/carpet a minimum of 100 square centimeters in size. This is placed in a wide-mouth polyethylene jar or zip-lock bag. At the laboratory, the sought after substance is extracted, suspended in water, and filtered through a polycarbonate filter or cellulose ester filter for analysis by transmission electron microscopy. This method is primarily used for asbestos analysis in carpet samples.

ANALYTICAL METHODOLOGIES

Depending on experience and equipment, the microscopist has the ability to detect, identify, and measure trace quantities of a substance down to the elemental composition and structural configuration of molecules. Most particles larger than

Table 14.2 Comparative Sampling Approaches for Asbestos-Contaminated Carpet Analyses

Sample Number	Carpet Piece	Microvac
1	4,800,000	21,000
2	3,300,000	30,000
3	<5,400	<350
4	3,800,000	74,000
5	3,000,000	50,000
6	2,500,000	95,000
7	3,600,000	18,000
8	4,700,000	35,000

< The less than symbol indicates the limit of detection for the method used.
Excerpted from *Methods for the Analysis of Carpet Samples for Asbestos. Environmental Choices.*[10]

1 micron in size can be identified by visible light microscope analysis. On the low end of sizing are some bacteria (e.g., 1 micron), many molds (e.g., 1 to 10 microns), and actinomycetes (e.g., 1 to 5 microns). On the high end are some molds (e.g., up to 50 microns in size), fibers (e.g., in excess of 100 microns in length), and hair (e.g., 10 to in excess of 100 microns in length). Then, too, particles may carry chemicals on their surfaces (e.g., formaldehyde adsorbed onto the surface of dust). These, too, may be identified.

Particles that are 1 micron, or less, are more difficult, if not impossible, by visible light microscope analyses. Under these conditions, special microanalytical procedures may solve the dilemma and/or confirm suspect materials. Forensic microscopy is only limited by the skills and experience of the microscopist. Some of the methods used are mentioned herein.

Visible Light Microscopy

An experienced microscopist may identify most, not all, sample particulates (greater than 1 micron in size) in a few seconds, if not immediately, without altering the chemical and physical properties of the material. Like differentiating a tree from a light post, when seen and identified on a frequent basis, most microscopic material is easily identified.

Visible Light Microscopy:
identification of particles greater than 1 micron in size

Parameters that are used in the microscopic analyses include, but are not limited to, the following:

- Size
- Shape
- Color
- Homogeneity
- Transparency
- Magnetic qualities
- Elasticity
- Specific gravity
- Refractive indices
- Birefringence
- Extinction
- Dispersion staining

Sample particles (larger than 1 micron in size) are identified with minimal effort by visible light microscopy. This is done through differences in size, shape, homogeneity, color, transparency, magnetic qualities, and specific gravity. Again, on occasion, additional information is necessary for positive identification of suspect material or for identification of a substance that is unfamiliar to the microscopist.

The remaining parameters are ascertained through the use of polarized light microscopy, which greatly increases particle characterization and competence. It also ensures positive identification of types of fibers, minerals, and some industrial pollutants.

Specialized Microscopic Techniques

Specialized techniques are available for use when the particle size is less than 1 micron, when the price is not a consideration, or for confirmation in litigation or high visibility cases. Occasionally, other techniques become necessary when dealing with extremely small particles or exotic mixtures.

X-ray Diffraction[11]

Other than visible light microscopy, X-ray diffraction is the only other technique available that permits identification and differentiation of crystals. Where there are three different forms of silica (i.e., quartz, tridymite, and cristobalite), chemical analysis may confirm the presence of silica but not be able to differentiate the type.

The sensitivity is down to approximately 10^{-2} nanograms, and the procedure is nondestructive of the sample. Although it may be as small as one particle, measuring 5 microns in diameter, the ideal sample size is 40 to 50 microns. The smaller particles require more extensive manipulation (e.g., removal of air from within the camera), so the cost for analysis may increase with smaller particle sizes.

X-ray diffraction measures the interplanar spacing of atoms in a crystal. Spacing is unique for every compound, and each is identified by comparison with known compounds. This comparison is performed with the assistance of a computer file that has well over 20,000 substances in its data bank. The data file is constantly being expanded. If a sample is suspect of containing a specific substance that is not on file, the known substance (in its pure form) may be scanned and entered into the data banks to be used as a reference for the unknown.

Scanning Electron Microscope[12]

Scanning electron microscopy (SEM) comes into play where a sample size is too small to be observed by visible light microscopy (equal to or less than 1 micron in diameter) or where greater resolution and depth of field of larger particles (from 1 to 100 microns in diameter) is required. It operates in a similar fashion to that of the stereo binocular microscope, refracting electron beams (instead of visible light) off the surface of a sample. These refracted electrons are projected onto a viewing camera or film to permit the analyst to observe the structure(s).

Scanning Electron Microscopy:
morphology, spacial, and inorganic elemental
analysis of particles down to 0.2 micron in size

The SEM is capable of magnification of particles typically around 0.2 micron in diameter. Where the depth of field for visible light microscopy is around 1 micron, it is 300 microns for the SEM. This allows for greater contrast and ease of viewing the unknown sample.

The resolution is about 300,000 times the actual particle size, or 200 times greater than that of the most powerful light microscope (which has a magnification capability of 1,500 times). Smaller particles are more readily observed due to the increased magnification. This is generally the case with metal fumes, clays, some pigments, bacteria, and viruses.

Another added feature to the SEM is the ability to add an energy dispersive X-ray analyzer (EDXRA) to the unit. This X-ray analyzer is capable of greater detection than that of X-ray diffraction. Where the X-ray diffraction provides a means for identifying compounds, the EDXRA can detect elements (above nitrogen on the periodic table). Most analysts agree that without this added ability for detecting elements, SEM would be inferior to visible light microscopy in its detection ability.

Still spores, bacteria, and viruses can be identified as spores, bacteria, and viruses only. Viruses and bacteria are generally smaller than 1 micron in diameter and can be identified as such only through the use of SEM due to the higher resolution. To type these biologicals by genus and species, the sample must still be cultured or manipulated in some other fashion besides microscopy.

Particles greater than 1 micron in diameter may still require the EDXRA for identification and are frequently more easily identified by visible light microscopy. Thus, SEM with EDXRA may be used as a secondary means of identification for the larger particles, and as the primary means of analysis for particles at or less than 1 micron in diameter.

Transmission Electron Microscope[13]

Transmission electron microscopy (TEM) analysis works in a similar fashion to that of the biological microscope by penetrating a sample with focused electron beams instead of visible light. These electron beams are observed in a similar fashion to that of SEM where the beams are projected onto a viewing screen or film.

> Transmission Electron Microscopy:
> identification and product analysis of particles/
> components down to 0.5×10^{-3} µm in size.

The depth of focus is 1 micron and its resolution is 0.5×10^{-3} micron. The maximum prepared particle thickness is 0.05 micron, and the maximum sample diameter is 3 millimeters. The TEM can be fitted for selected area electron diffraction (SAED) and EDXRA. The SAED functions in a similar fashion to that of X-ray diffraction, limiting the coverage area, and the electron beam is used to measure the interplanar spacing of atoms in a given area. The SAED information is compared with a data bank for compound identification, and the EDXRA provides the elemental fingerprint.

Particles are scanned for structural appearance, compound identification, and elemental fingerprinting. If a search is performed for a specific substance, the microscopist reports results in percent by weight. It is important to note that where asbestos content is performed by polarized light microscopy, the results are provided as percent by volume where the definition of asbestos is a mixture containing greater than 1 percent by weight of certain types. The only means available to provide true percent by weight is through TEM.

Electron Microprobe Analyzer[14]

The electron microprobe analyzer (EMA) is an ultra-micro-analytical tool that can be used to enhance a light microscope, SEM, X-ray fluorescence, and cathode luminescence. It is also referred to as mass scanning.

A sample containing a large number of small particles may be rapidly characterized by chemical composition. This is generally performed by automation of the specimen stage, scanning beam, and spectrometer. In this case, a few thousand particles can be characterized from any given sample.

The electron microprobe analyzer may be used to locate a needle in a haystack. If searching for a known substance which may be present in a sample in only parts per million, or trace levels (e.g., asbestos fibers in urban air), the analyzer is ideal. It is set up to identify an element, or combination of elements, which are present in the substance of concern. Each time the substance is located, the stage stops, and that particle is quantitatively analyzed. Then, the stage continues its search. The ideal lower limit for adequate identification is 0.1 percent, but the method is capable of locating down to 10^{-4} percent. The latter involves a considerable amount of time consumption, therefore, cost, but it is possible.

Samples as small as 1 micron in diameter can be analyzed for most elements present in the sample to 1 percent or greater. This constitutes a detection limit as low as 10^{-4} nanograms. An electron beam is focused to a spot smaller than 1 microns square in area. The characteristic X-rays emitted from the spot are analyzed for wavelength, or energy, dispersing systems both quantitatively and qualitatively. A limitation is its inability to detect lithium (sometimes beryllium), and one hundred parts per million is the lower limit of detection for most elements.

The analyzer is capable of mass scanning of between 4 and 50 particles per hour. The speed and ease of analysis allows for any given sample, containing up to 1,000 unknowns as little as 20 hours to analyze—a 24-hour turnaround, in a pinch!

Ion Microprobe Analyzer[15]

The ion microprobe analyzer (IMA) provides a means for mass spectrometry on small particles or small areas of bulk samples. This method is one of the most sensitive tools available for small particle analysis. It is sensitive to every element in the periodic table and can, under ideal conditions, detect as little as 10^{-20} grams of some elements, and 10^{-19} grams of most elements. It is fully capable of analysis of trace amounts of material from samples as small as 1 micron and, in some instances, can obtain parts per billion of some elements. The time required for semi-quantitative analysis is typically 40 seconds, or as short as 4 seconds.

The instrumentation consists of a light microscope (which is used to locate the sample), an ion source, a column of two electrostatic lenses, and a mass spectrometer.

The versatility of the IMA is similar to that of the electron microprobe analyzer, yet it is much faster and can assess particles which are much smaller (e.g., 1 part per billion instead of 1 part per million). Any airborne, waterborne, or contaminant particles can be analyzed with this tool. A few examples include the analysis of micrometeorites, lead particles from auto exhaust, and contaminants on integrated circuits.

COMMERCIAL LABORATORIES

Due to the extensive training required to be proficient in all aspects of this field, there are a limited number of laboratories capable of responding to all the nuances that may arise in an environmental evaluation. An experienced microscopist may cost a little more per hour yet be able to provide results with less time expenditure than one with less experience and lower rates.

On the other hand, the desired information may be obvious (e.g., heavy concentrations of ragweed pollen) and readily apparent to even the inexperienced microscopist who may serve as an initial pre-viewer. Many of the commercial laboratories are accustomed to analyzing primarily for asbestos only. Quarry the commercial laboratory as to its capabilities and limitations. Those experienced in performing forensic analyses can easily apply the forensic knowledge to environmental issues.

Charging only a nominal fee to perform a more complete analysis, some mold/spore laboratories provide results in terms of total fibers. A forensic laboratory should be able to identify each of the fibers (e.g., rodent hairs, fiberglass, cellulose) and much more.

SUMMARY

In conjunction with an experienced laboratory, forensic dust sampling is a very powerful tool for identifying a broad range of unknowns. Not only can those substances which are more commonly encountered be identified, but forensic microscopy provides an avenue for identifying unknowns where there are no traditional methodologies or no readily available means for analysis. Not only can particles be identified, but many metals and chemicals can be identified. This approach to indoor air quality takes the investigator to a new dimension in sampling methodologies.

REFERENCES

1. McCrone, Walter C. Detection and Measurement with the Microscope. *American Laboratory.* [Reprint] December 1972.
2. Hedge, A., W. Erickson, and G. Rubin "Effects of Man-Made Mineral Fibers in Settled Dust on Sick Building Syndrome in Air-Conditioned Offices." Proceedings from a Conference on Indoor Air, 1993.
3. McCrone, Walter C. The Solids We Breathe. *Industrial Research.* April 1977.
4. Bisbing, Richard. Clues in the Dust. *American Laboratory*. [Reprint] November 1989.
5. McCrone, Walter C. Microscopy and Pollution Analysis. Reprint from "Measuring, Monitoring, and Surveillance of Air Pollution," *Air Pollution*. Volume III (1976).

6. McCrone, Walter C. *Air Pollution.* Academic Press, Inc., New York, 3rd Edition, Volume III, 1976. pp. 101-2.
7. Bisbing, Richard E. Microscope and Pollution Analysis. [Oral communication] McCrone Associates, Inc., Chicago, IL. (June 1995).
8. Millette, J.R., T. Kremer, and R.Wheeles. Settled Dust Analysis Used in Assessment of Buildings Containing Asbestos. [Bulletin] McCrone Environmental Services, Inc., Norcross, Georgia, 1990. pp. 216-219.
9. Millipore Product Literature. Millipore Corporation, Bedford, Massachusetts, 1996.
10. Millette, James R., et. al. Methods for the Analysis of Carpet Samples for Asbestos. Environmental Choices Technical Supplement, March/ April 1993.
11. McCrone, Walter C. *Air Pollution.* Academic Press, Inc., New York, 3rd Edition, Volume III, 1976. pp. 114-115.
12. Ibid. pp. 118-121.
13. McCrone, Walter C. *Air Pollution.* Academic Press, Inc., New York, 3rd Edition, Volume III, 1976. pp. 121-132.
14. Ibid. pp.132-138.
15. Ibid. pp.138-143.

Chapter 15

ANIMAL ALLERGENIC DUST

The neglected partner in allergenic complicity with pollen and mold spores is animal allergens, or house dust. Only within the past ten years have clinical studies revealed a strong relationship between levels of animal allergens in dust and allergy symptoms. Technology has evolved. Methods have been refined, and immunoassay technology comes into the limelight.

Detection and quantitation of a wide range of antigenic biological and nonbiological substances are now possible through immunoassay analytical methods. Allergenic substances that are processed include proteins, glycoproteins, hormones, peptides, chemical haptens, and drugs. Of particular interest to the environmental professional, researchers have developed immunoassays for animal allergens, predominantly those derived from mites, cats, cockroaches, and rodents. Methods have also been developed for certain species of fungi (e.g., *Aspergillus flavis*) and for latex.

An immunoassay involves identification of the antigens by creating antibodies for the express purpose of tagging specific materials. Quantitation is based on the antibody-antigen complexes. Thus, immunoassay analyses are highly specific and quantifiable.

The sampling procedure is simple and inexpensive. Yet, the sampling strategy and results interpretation require a thorough understanding of the process.

In the past, the medical community has performed the sampling, but most of the sampling has been diagnostic, involving expensive clinical tests performed on the distressed sufferer. Where allergies appear widespread in an office building or other problematic indoor air environment, the perplexed facilities manager or homeowner seeks assistance from the environmental professional. Diagnostic tests on all the building occupants can be expensive and time consuming. In such instances, dust sampling is by far the more feasible alternative.

Yet, without the benefit of clinical studies, extensive allergy complaints may pose a medley of possibilities. Some of the allergy sufferers know the specific antigens which cause their individual reactions. Known allergens may assist in narrowing the possibilities, where dust sampling may be performed in order to:

- Define allergen levels in residences of asthma patients.
- Identify areas or sources of elevated levels of allergen(s).
- Determine the effectiveness of allergenic-dust control measures.

Most of the information provided within this chapter is intended to aid in the search for the more common allergens and expound on those that may be overlooked in isolated instances. The animal allergens are more widely understood and are an evolving issue of concern in the environmental field.

ANIMAL ALLERGENS

Animal proteins are high molecular weight, complex molecules which can illicit an allergic reaction. Wherein the environmental professional is concerned with airborne exposures, the allergens must be present in large quantities and small enough to become airborne. Typically, those animal allergens that are more commonly encountered are parts-and-pieces of an insect or mammal. Those that receive the greatest attention and are frequently studied are dust mites, dog/cat dander, and cockroach body parts. The probability of elevated numbers of these allergens is considerable in most indoor air environments.

Mites/Spiders[1,2]

Mites are small to microscopic-sized, generally parasitic arachnids with four pairs of legs in their adult stage and little or no differentiation of the body parts. Many cause allergic rhinitis, human dermatitis, and general allergic reactions. They differ in habitat and associations and are broadly categorized by their associations. See Table 15.1 for a breakdown of the most commonly cited allergenic mite types.

Storage mites are usually of the genera *Lepidoglyphus* and *Tyrophagus*. They rely on decaying vegetation as a food source and are normally associated with agricultural environments. They have been identified as causing allergic rhinitis in dairy farmers. Thus, allergy-causing storage mite exposures are limited more to the outdoor environments where there is decaying vegetation than to the indoor air environment. Decaying vegetation is a requirement for their presence.

Itch mites and mange mites may cause a dermatitis and have occasionally been implicated with house dust mites. Their main food sources are cheese, dried meats, flour, and seeds. They damage and contaminate these commodities while having a means of transportation through human handling of the food products. The result is grocer's itch, or miller's itch.

Some mites and spiders attack man and other animals directly, burrowing into their skin. These are the ones which hikers and hunters often encounter in wooded areas.

Others cause mange, which results in itching and hair loss. Mange mites typically attack domestic animals and are generally visible to the naked eye. Yet, following a heavy infestation, dead or alive, their bodies may still serve as antigens.

House dust mites have undergone considerable study as they are not only allergenic, but they typically are found indoors. See Figure 15.1. They bask in warm, moist, dark environments. Ideal temperatures are between 70 and 80° F. They thrive where the relative humidity is in excess of 65 percent, and they hide from sunlight.

Sites where they tend to commune are places that have sluffed epithelial cells (which tend to retain moisture), such as beds, upholstered furnishings, and carpets. The average human will lose as much as five grams of epithelial skin cells per week. Wherever these epithelial cells can be found, the mites have a source of food.

Table 15.1 Allergenic Mites

Class: Arachnida Order: Acari (Acarina) Suborder: Psoroptoidea	
STORAGE MITES	
Family: Acaridae, Glycophagidae, and Blomia	
Genus/species:	*Acarus siro* *Glycophagus domesticus* *Lepidophagus destructor* *Tyrophagus putrescentiae* *Blomia tropicalis*
ITCH MITE	
Family: Sarcoptidae	
Genus/species:	*Sarcoptes scabiei*
DUST MITES	
Family: Pyroglyphidae	
Genus/species:	*Dermatophagoides pteronyssinus* *Dermatophagoides farinae* *Dermatophagoides microceras* *Euroglyphus maynei*

Excerpted from *Allergy Basics for IAQ Investigations.*[3]

In the United States, dust mite infestations and allergies tend to be seasonal with a preference for the warmer, humid months (e.g., summer). Whereas tropical climates may provide a perpetual, unrestricted habitat for blissful invaders all year round, the pesky little critters predominate mostly between June and August.[4]

The house dust mites are between 250 and 500 microns in size, barely visible by the naked eye and frequently overlooked. The allergenic portion of these mites is thought to be the body parts-and-pieces and their fecal material, which is 10 to 35 microns in diameter. The body pieces are considerably smaller than 250 microns and not identifiable by microscopic analysis. As a matter of course, the size has a bearing on the airborne allergens. If greater than 40 microns, airborne substances

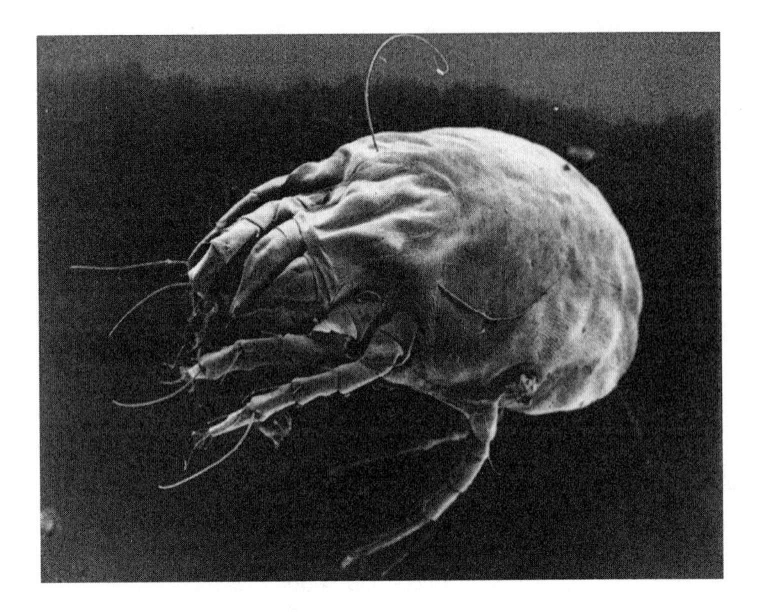

Figure 15.1 The dust mite is a commonly used representation of allergens. (Courtesy of ALK-Abelló A/S Hønsholm, Denmark.)

will settle out within 20 to 30 minutes. Thus, inhalation of the material is most likely to those components of the dust that are smallest and in areas often disturbed.

Areas likely to be disturbed are situation dependent. Commonly involved activities that might stir up dust include, but are not be limited to, the following:

- During and shortly after vacuuming
- Considerable activity on and disturbance of upholstered furniture
- Considerable activity on and disturbance of carpeting
- When making a bed and fluffing pillows
- Sleeping on contaminated bedding and/or upholstered furniture

As for the actual allergens, several mite-associated proteins are implicated, and studies have been predominately on three species in the genera *Dermatophagoides*, which is common in North America and Europe. To a lesser extent, the genera *Euroglyphus* and *Blomia* have been studied as well, but they are more commonly encountered in Central and South America.

For the purpose of allergen testing using immunoassay techniques, the dust mite allergens are genus and species specific, and each of the species has as many as 40 different proteins that could cause an allergic reaction. Where identified, allergenic proteins are referred to by group. See Table 15.2 for the most commonly implicated allergen types and associated allergenic proteins.

Table 15.2 Allergenic Dust Mites and Immunoassay Test Groupings

Type	Allergenic Proteins Group I	Group II	Group III
Dermatophagoides farinae	*Der f* I	*Der f* II	*Der f* III
Dermatophagoides pteronyssinus	*Der p* I	*Der p* II	*Der p* III
Dermatophagoides microceras	*Der m* I	—	—
Euroglyphus maynei	*Eur m* I	—	—

Where a group of allergenic proteins has not been identified, a homogeneous mix is referred to as polyclonal. A polyclonal assay involves multiple antigens from the same life form.

Booklice

Oftentimes, the layman will refer to paper mites as being the source of a problem, possibly because the individual associates their allergies with the mounds of paper they work with and street hearsay. The reference to paper mites is a red herring, a fictitious contrivance of the news media. Entomologists frown in a desperate attempt to track these illusive pests under the heading of mites. While some entomologists will confess ignorance, others will speculate that the reference is more likely to that of storage mites which, at times, are associated with cellulose, or paper products. Another consideration is that of booklice are neither mites nor lice, but insects.

Booklice belong to the order Psocoptera. These are small, soft-bodied insects with three pairs of legs and measuring less than ¼ inch in length. They may or may not have wings. They have been reported as causing allergic symptoms in places with large amounts of paper. They feed on molds, fungi, cereals, pollen, and dead insects. Their preferred habitat is moist areas and humid environments, and they rarely cause damage to the spaces they occupy. They are, however, a nuisance to allergy sufferers.

Cockroaches and Other Insects[5]

Insects are typically visible, have three pairs of legs in their adult stage, and possess three distinct body regions. They are, therefore, easy to identify, and a large indoor insect population rarely remains unnoticed. The most common is the ever-present cockroach.

Of the 55 species of cockroaches that inhabit the continental United States, less than ten are indoor residents. Of these, the most common, particularly in southeast-

Figure 15.2 Photomicrograph of an "unconfirmed" booklice, a component of office dust, lifted by tape. Its approximate size is 500 microns, and this image was observed, in its entirety, under 100x magnification. Immunoassay testing was for cockroach allergens only, which were found to be excessive. Dust mite allergens were low, but the method was specific for dust mites, not booklice.

ern United States areas, are the larger American cockroach (*Periplaneta americana*) and the smaller German cockroach (*Blatella germanica*). See Figure 15.3. As the larger ones consume the smaller, more prolific ones, they do not tend to cohabit within the same residence. It should be visually apparent as to which species one is dealing with. See Table 15.3 for the allergenic groups.

Recent studies, however, suggest that the cockroaches secrete their allergens onto their bodies and other surfaces in their environment. Thus, examination of allergenic material may or may not disclose the presence of associated debris and fecal material. The only means of confirming the presence of cockroach allergens is through immunoassay analysis of suspect dust.

Table 15.3 Allergenic Cockroach Material and Immunoassay Test Groupings

	Allergenic Proteins	
	Group I	Group II
Blatella germanica	*Bla g* I	*Bla g* II
Periplaneta americana	*Per a* I	—

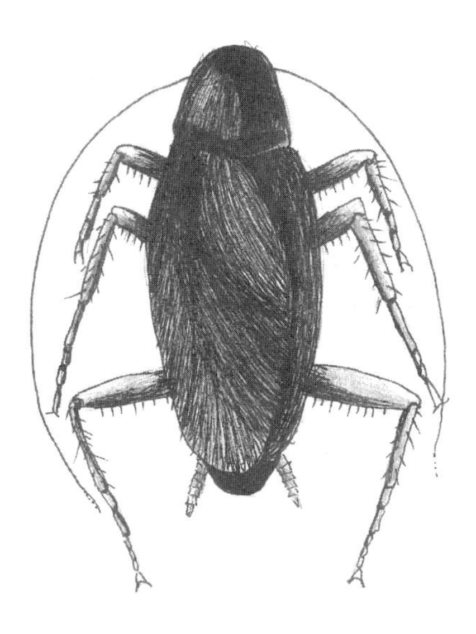

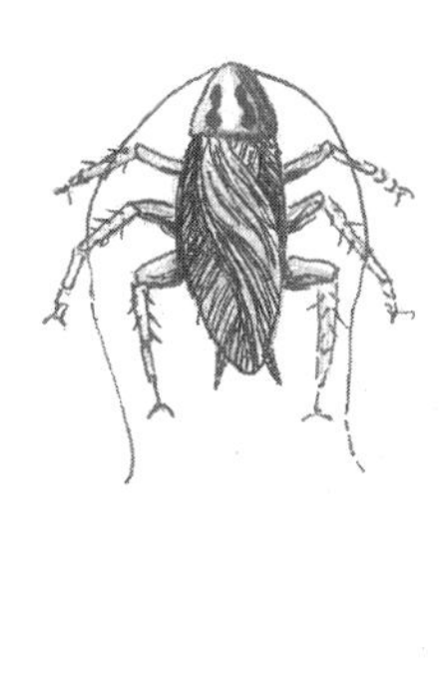

Figure 15.3 The American cockroach (*Periplaneta americana*) [left] and the German cockroach (*Blatella germanica*) [right].

There has been considerable, unconfirmed speculation as to the source of the cockroach allergens. Some considerations are as follows:

- Saliva
- Body parts-and-pieces
- Egg shells
- Fecal particles

Cockroaches are able to adapt to low ambient humidity, yet actively seek a source of water. For this reason, indoor cockroaches are most likely found around water pipes, pet water bowls, evaporative areas around refrigerators, leaking faucets, and wet carpeting. Although in most cases their presence is readily apparent, cockroach allergens have been measurable in up to 15 percent of homes which had no visible clues that they might be present. Keep in mind that they do not have to be alive for the allergens to promote a reaction, and the source of cockroach allergens is still unclear.

Although there have been numerous allergen studies performed of the ever-enduring, ever-present cockroach, many other insects have been implicated as well. Though most other insects are generally found in outdoor environments, body parts-and-pieces may be conveyed indoors. They may attach to clothing. They may enter open windows and doors in search of food or light (e.g., June bugs). There may be air movement from the outside to enclosed air spaces indoors. Indoor accumulations of insect carcasses are common. The indoor environment may become a repository of debris.

Outdoor workers are exposed, at times, to insect fragments and debris at levels in excess of ragweed pollen. Insects that are suspect of causing allergies include the following.[3]

- Crickets
- Houseflies and fruit flies
- Waterfleas
- Bed bugs
- Mayflies
- Aphids
- Honey bees
- June bugs
- Bean weevils
- Some species of moths
- Butterflies
- Silkworms
- Caddis flies
- Chiromomid midgets

Some occupations, due to their associations, harbor potential exposures for insect infestations. Some examples may be found in Table 15.4.

Table 15.4 Occupational Exposures to Insects

Entomologists: locusts, crickets, flies
Grain mill workers: beetles, grain weevils
Loggers/lumber mill workers: Tussock moth
Fishermen/bait handlers: meal worms, maggots
Poultry workers: Northern fowl mites
Bakers: storage mites, grain weevils
Small animal handlers: fleas
Honey-packing plant workers: honey bee dust
Pet food processors: Chironomids

Domestic Animals[6]

Most allergic substances that are associated with domestic animals are predominately exposure problems in environments where the animals reside. The obvious is overstated. Homes, kennels, pet shops, and laboratories are locations where the allergens are most likely to be found. However, office environments should not be excluded from consideration. Although rare, elevated exposures have been reported where the source has not be readily apparent. The most commonly targeted domestic animals are cats and dogs.

Cats

Although cats are maintained in 28 percent of all American households, only two percent of the U.S. population has allergies to them. Interestingly, those who are allergic to cat allergens may never have lived with cats.

The source of allergenic material may be any of a number of feline-associated materials, and tests are performed for *Felis domesticus* (*Fel d* I). There has been considerable speculation as to the actual chemistry of the allergen, but most researchers have speculated that the allergen is somehow transferred, picked-up, and/or concentrated by saliva. When the cat grooms itself by licking, the allergen is spread or transferred to the hair and epithelial cells. The following is an abbreviated list of known and suspect sources/transfer vehicles:

- Saliva
- Sebaceous glands
- Hair
- Epithelial cells
- Epidermis

The sex and type of cat are factors which contribute to environmental levels of cat allergen. Male cats shed more allergens than female cats. Patient's symptoms vary in severity, depending upon the type of cat to which they are exposed to (e.g., a domestic cat versus a Persian short-tail).

The cat allergens are carried on particles less than 2.5 microns in diameter. At this size, after becoming airborne, they will remain suspended in an undisturbed environment for hours. The ease with which these allergens become airborne is the reason for apparent excesses in sensitivity.

Even though there are excessive cat allergens in a household where one resides, all indoor environments have detectable levels. In houses with cats, there is typically in excess of 10 micrograms per gram (μg/g) of *Fel d* I in the dust, and levels have been reported as high as 7,000 μg/g.[7] The allergen may also be transferred by clothing and other articles from a high exposure environment to otherwise cat-free environments. Houses that have never had cats may have levels of less than 1 μg/g in the dust, but levels in excess of this should not be surprising.

Dogs

Dogs are maintained in an estimated 43 percent of American homes, and one regional study indicated that as many as 17 percent of the population was allergic to dog allergens. There have been twenty-eight different allergens found to be associated with allergic symptoms. The specific antigen to which most patients react is designated as *Can f* I. *Can f* I consists of an extract of the following dog-associated materials:

- Hair
- Dander
- Saliva

Most environments where dogs are found have in excess of 120 µg/g. of *Can f* I in the dust. Homes without dogs typically have less than 10 µg/g. The size of the allergen-carrying material or contaminated particles is unknown.

Rodents[8,9]

Exposures to mouse and rat allergens are typically associated with animal research laboratory vivarian and indoor spaces (e.g., homes and office buildings) infested by rodents. Of the approximately 35,000 workers in the United States exposed to rodent allergens in animal research laboratories or breeding facilities, over 20 percent of the workers experience allergic symptoms.

Rodent infestations may deposit allergens unbeknownst to building occupants. Awareness of the potential opens another door for search and disclosure of possibilities. Complaints of a urine-like odor should alert suspicion.

Two other allergens have been associated with mice. One of the mouse allergens, referred to as Ag1, is related to the mouse urine and is designated *Mus m* I. *Mus m* I is produced by the liver and salivary glands, excreted in the form of urine and saliva. As it is associated with testosterone, *Mus m* 1 is excreted predominately by male mice. Other factors affecting quantity are strain and age. The other mouse allergen, referred to as Ag3, has been detected in hair follicles (e.g., fur and dander extracts). This is designated Ag3.

Two allergens have also been associated with rats. However, they are both associated with the rat urine. These allergens, referred to as Ag4 and Ag13, are designated *Rat n* I.

Although typically associated with particles of 10 microns in size or less, airborne exposures to rodent urine rarely occur unless contaminated bedding is disturbed, and elevated humidity has been reported to diminish airborne exposures. Most exposures occur where rodents are maintained in large numbers (e.g., laboratory environments). The amount of material that may become airborne is generally related to the type of litter and bedding. The allergen is generally released into the air during cage-cleaning activities. In one laboratory animal cage-cleaning study, the airborne levels were reported between 19 and 310 nanograms per cubic meter of sample (ng/m^3). During quiet times, when there were no disturbances, the levels dropped to around 1.5 to 9.7 ng/m^3. Rodent allergens may also be deposited on ceiling tiles and in carpeting. Disturbances of contaminated areas may result in airborne releases of material as yet not identified to be present in a given environment (e.g., rodent infestations). There is no published information regarding reported airborne or dust levels of rodent allergens.

Farm Animals[10]

Along with the arthropods, pollen, mold spores, and bacteria, farmers and farm workers are potentially exposed to farm animal allergens. The more prominent allergenic exposures are attributed to cows, horses, and pigs.

Cow dander and urine, designated *Bos d* II, have been reported to cause allergic rhinitis in dairy farmers. Airborne levels have been reported as high as 19.8 μg/m^3.

Horse allergens (*Equ c* I, *Equ c* II, and *Equ c* III) are very potent. Exposures may occur occupationally or to pleasure horseback riders. The horse allergens are related to hair, dander, and epithelial cells.

Pig allergens are rarely reported to be a problem and are considered weak antigens. The allergenic material has, however, been identified. Swine workers have been found to have antibodies against swine dander, epithelium, and urine. Airborne levels have been reported up to 300 μg/m^3. Yet, there appear to be minimal complaints and concerns for swine allergies.

Other Animals[10]

Rabbit dander and guinea pig urinary proteins/saliva have been reported to cause allergies. Both are found in homes, pet stores, and laboratory facilities.

Even rarer are exposures to bat guano and reindeer epithelial cells. Bat droppings accumulate inside roof attics and cave dwellings. Asthma-like symptoms are generally reported in association with workers exposed to bats in indoor working environments.

Reindeer epithelial cells, which are associated with leather processing are also known to cause allergic reactions. Airborne exposure levels have been reported in a workshop at concentrations of 0.1 to 3.9 μg/m^3.

OCCURRENCE OF ANIMAL ALLERGENS

Farm, laboratory, and pet environments are easy marks. The source of allergies is direct and readily apparent to the allergy sufferer when symptoms worsen in their presence. The greater the number of animals, the greater the potential for elevated exposures. Whereas an individual may not at one time have been sensitive to a given allergen, an extreme dose may later predispose them to developing symptoms at lower exposure levels in the future. Most of these allergy sufferers know what it is they are allergic to. A small percentage of the population, however, seems to be sensitive not only to the typical allergens, but to just about everything.

In office environments with no apparent sources of animal allergens, the latter more allergen-sensitive individuals will be the first to start complaining. It has been estimated that these ultrasensitive individuals constitute only about 4 percent of the population. Yet, as levels of an allergen increase, the numbers impacted increase as well. With more complaints comes greater concern for locating a source. As animal allergens are possible contributors to a given environmental invasion, a means for identifying and quantifying their presence is made available through a well thought-out strategy, collecting dust samples, and analyzing the collected material by immunochemisty.

SAMPLING STRATEGY

A well thought-out strategy is vital for identifying a problem and obtaining meaningful results. If the environment is an office building and large numbers of people are impacted, the problem areas must be clearly identified.

Questionnaires should be filled out by all those in the area of concern as well as an area where there are no complaints of allergy-like symptoms. Identify known problem areas and non-problem areas. Develop associations. Attempt to limit the possibilities. A screening tool for rodents, using ultraviolet light, may also be added to the list of considerations. The method is discussed in the next section of this chapter.

Other than bacterial and mold spores, the most typical allergenic materials found in office buildings are dust mites and cockroach allergens. Cat and dog allergens are more common in homes but may be transferred to office environments. All other allergens previously mentioned are rare occurrences in office/industrial environments or are occupationally related.

Some building occupants may have had allergy testing (e.g., skin or allergy blood tests) performed and know what allergens to which they are allergic. A few of the more common blood tests available through physicians include the following:[11]

- Plant pollen
- Fungi/molds
- Thermophilic actinomycetes
- Storage mites
- Animal dander
- Isocyanates
- Formaldehyde
- Gums/adhesives
- Anhydrides

The greatest dilemma to the environmental professional is that of locating areas most likely to be source origins and including only these in the sample, not the disassociated areas as well. For instance, if complaints generally arise in a given area where there is a lot of disturbance of the carpeting and dust mites are suspect, the area where the traffic passes should be sampled to include as much of the known problem material as possible, including areas under desks and along walls which may or may not be the source origins. Each falls within a different source type (e.g., carpeting) and functional grouping.

Sample sites should be selected based on suspect source types. These may include, but not be limited to, the following locations:

- Carpeting
- Upholstered furniture
- On top of ceiling tiles
- Ducting in an air handling unit

Functional grouping of sample sites is the most difficult to identify. The impacting function may or may not be obvious. In most cases, it will not be obvious, and several functional areas will require sampling. Grouping may include, but not be limited to, the following:

- Dusting shelves
- Vacuuming the carpeting
- Excessive traffic
- Maintenance involving removal of ceiling tiles
- Air handler activity

The number of samples taken will depend upon the environmental professional's assessment of the situation. This will vary in a case by case situation, depending upon the number of possibilities ascertained to be a potential problem source. Then, too, sampling of a non-problem area may be desirable for comparative purposes.

During data/information gathering, clarify whether a suspect carpet was recently shampooed or vacuumed. Maybe there has been a recent infestation of cockroaches or rodents. Even where the vermin have since been exterminated, their body parts-and-pieces may be the exposure allergens. In locating possible sample sites where these parts-and-pieces may have been deposited, the environmental professional should also consider the function of that location as well. For instance, rodents may eat through air supply ducting and leave a trail of urine and feces. Allergenic material deposited in an air plenum will pose a greater potential for occupant exposures than the same material deposited in the corner of a room where there is no foot traffic or air movement.

Record the sample area size and exact location. Although the area size may not be relevant (samples are analyzed by weight comparisons), this additional information may be useful at some future date, and some professionals do standardize the sample size (e.g., one cubic meter of area). You, once again, are not obligated to do the same.

Although there is no set protocol, sample sites should be clearly identified. If floor plans are available, indicate the limits of each site and assign a sample number to the area. Otherwise, describe in detail the specific location(s) (e.g., on top of the ceiling tile above the copy machine) with its perceived function (e.g., frequent above ceiling work necessitates disturbance of ceiling tile at this location).

SCREENING FOR RODENTS[12]

Public health food inspectors screen for the presence of rodents in a food processing plant by using ultraviolet light. Under ultraviolet light, urine will fluoresce bluish white to a yellow white. Fresh stains fluoresce blue while the older stains shift to a more yellow color. Rodents tend to urinate while in motion, thus leaving a characteristic droplet trail. Rodent hairs will also fluoresce bluish white and can be easily identified in areas where they hang out (e.g., food storage areas).

This characteristic of urine allows visual inspection for tale-tell signs in dimly lighted areas. The darker the area under investigation, the more clearly visible will be the fluorescent stains.

SAMPLING METHODOLOGIES

Due to the complexity and lack of clinical comparison studies, sampling is typically performed on settled dust. Even though air sampling can as easily be performed, airborne levels vary considerably, because they are based on dust generating activities that are in progress during the sampling period. Depending upon the airborne dust levels, the required sample air volume may be in excess of 5,000 liters. Large air volumes, in turn, embrace extended sampling times and/or environmental air sampling devices that are capable of drawing 100 to 1,000 liters per minute. If sampling times are extended, the activity that generates airborne dust may not be singularly represented in the sample. Then, environmental sampling equipment is expensive and cumbersome. There are no published procedures for allergenic dust air sampling, and researchers shy away from this approach. On the other hand, settled dust sampling is simple and clinical comparison studies have been performed.

Settled dust sampling is as easy as sucking suspect dust into a vacuum cleaner bag or as involved as using specialty sampling devices designed to perform allergen dust sampling. The following sample collection devices have been successfully used by environmental professionals and allergists:[13]

Standard vacuum cleaner with a filter bag – The filter bag is later detached and sent to the laboratory for analysis. Studies have indicated that allergenic material thought to be retained in the hose preceding the bag should not be a significant concern to the sampler. They did, however, find that high retention, high efficiency bags do retain material better than the regular bags. Losses may be as much as 30 to 50 percent with the low retention bags. These may be purchased at some grocery/discount stores.

Commercial high-efficiency particulate (HEPA) vacuum cleaner with a high efficiency filter bag – The filter bag is later detached and sent to the laboratory for analysis. This apparatus is more efficient for retaining particles smaller than 2.5 microns in diameter than most conventional vacuum cleaners. To avoid cross-contamination, assure the hose and all connectors between the intake wand and filter have been cleaned prior to each collection.

Specially designed vacuum cleaner with an external filter attachment – One design incorporates a small bag that is inserted between the hose and the wand connection. Another uses a molded plastic wand that contains a filter.

Air sampling pump with a polycarbonate membrane filter cassette – Samples are drawn through the filter with an air sampling pump in much the same fashion as a vacuum cleaner (disregarding the flow rate) and the dust is collected directly from the surface area in question.

A private laboratory performed comparison tests for some of the above sampling devices and concluded that there is a significant difference in their allergen dust recovery. They ranged from half the original sample dose to double. For this reason, the same method should be used consistently, and comparison sampling of problem and non-problem areas is strongly indicated so as not to rely on threshold values only.[13]

Commercial laboratories emphasize that the principal consideration in sampling should be the quantity of material captured. Although some researchers propose sampling within a well defined, delineated area or a limited sample duration, the quantity of material collected may not provide a good representation of the environment or allow for a sufficient amount of collected dust to retain the desired sensitivity (in the picogram range). Those who define the area of surface coverage generally opt for one square meter. Others sample for a specified time period (e.g., two minutes), no matter what the substrate. They typically vacuum for a set period of time (e.g., 5 or 10 minutes), without regard to the area covered. Yet, in both instances, the amount of dust collected still relies on the amount of available dust. There is no such thing as too much collected dust. There may, however, not be enough.

Ideally, the environmental professional should be able to estimate the amount of dust collected and target a collection in terms of milligrams (with the volume dependent upon the density of the collected material). Some laboratories specify 200 milligrams. Others specify 500 milligrams. The latter allows for a certain fudge factor with plenty to spare.

Although not required, composite sampling is recommended where several samples are to be taken and the primary allergenic reservoir has not been identified. One study indicated that three composite samples taken from the same dwelling, each a week apart, gave similar results. There was, however, a noted difference between areas within the same dwelling and between dwellings. Thus, composite samples provide a relatively consistent estimate, and discrete samples were useful in finding specific reservoirs.[13,14] These samples may be taken during the collection process or involve a contribution of dust from each of the samples taken in areas known to be associated with airborne allergens. This approach can be useful for screening and minimizing the number of samples requiring analyses for all allergenic dust.

In brief, the most relevant consideration is locating the sample site. Isolate and identify a specific, suspect sample site (e.g., around an area known to be associated with allergic reactions) along with its function/activity (e.g., high foot-traffic area or bedding). Then, collect a sample (or a composite of several samples). Compare suspect problem sites with known non-problem sites. If an allergen appears suspect, discrete area sampling and analyses will aid in the identification of reservoir(s). Comparison sampling, coupled with published thresholds, will result in manageable interpretations.

ANALYTICAL METHODLOGIES

For the purpose of extending the reader's knowledge into the realm of understanding human test results and their applicability to a site investigation, human testing methods are discussed in brief within this section along with sample analytical methods. They are both relevant in assessing suspect allergens.

Human Testing

Airborne exposures to animal antigens may result in allergic rhinitis, sinusitis, and asthma. Although dermatitis and urticaria have not been implicated with airborne exposures, where an allergic individual rubs up against dust laden with a given antigen, the skin may be impacted as well.

Allergenic individuals may be tested by any of a number of means, each associated with a different route of entry or means of exposure. The simplest and most frequently used method is the direct skin test.

The skin test technique primarily allows for identification of direct contact allergens only. It will not provide for identification of airborne allergens. A suspect antigen or group of antigens are placed on the skin surface of the individual, usually a site on the arm, and a retentive barrier is placed over the material. The site is checked for redness and urticaria after 24 hours. A positive reaction indicates the individual is sensitive to all or one of the challenge allergens.

A less commonly used technique is that of blood sample analysis. Blood is extracted from the individual and analyzed by immunochemical techniques. This method is by far the less invasive, not challenging the individual allergy sufferer.

The immunochemical techniques used are the same as those used for the dust samples and are discussed in a little more detail under Allergenic Dust Testing.

The most ideal (yet impractical) technique is that of a direct bronchial challenge to the individual. This method provides a direct insult to the individual while they are restricted to a challenge chamber where the airborne allergen types and amounts may be controlled. Whereas one may not respond to the skin test, the bronchial insult chamber may elicit a respiratory reaction. The technique is mostly of the research mode and, if accessible, is very expensive and time consuming.

Allergenic Dust Testing

Commercial laboratories currently offer routine testing for dust mite, cockroach, cat, and dog allergens. Although an occasional commercial lab may be willing to extend their limits, several research institutions have the materials necessary to test for the less common allergens (e.g., *Mus m* I). Immunoassay testing of allergenic dust is extremely sensitive and highly specific. This specificity is advantageous in most cases but can be a drawback where a similar, but not identical, allergen is suspect.

Sample preparation involves sieving the dust samples to separate all material less than 300 microns in size from the larger material. After it has been weighed, the smaller material is then extracted with a special buffering solution. An aliquot is taken of this extract and analyzed by any of a number of immunoassay approaches, involving single antigen-specific or multiple antigen quantitation.

A single specific protein, or antigen, is referred to as "monoclonal." An example of a monoclonal test antigen is *Fel d* I. Being related to and being the strongest (or most studied) allergenic component of a given species, the monoclonal antigens become the most commonly sought after test material. Yet, due to the extensive amount of attention given to a limited number of allergenic species, the specificity of these methods may restrict the possibilities.

Those species that have received the most attention are also those that are known to cause allergies and are prevalent in significant numbers. Figure 15.4 shows the results of a study involving school children and prevalence of allergies. As a child ages, allergies diminish until ages 25 to 34.[15] Thus, the prevalence of the various allergies is likely to be less in the adult population.

Multiple proteins from one species or several are referred to as "polyclonal." An example of a polyclonal test antigen is cat allergens. They have not been as extensively studied as the monoclonal antigens, and results seem to be less consistent and are more difficult to evaluate.

The identification of polyclonal antigens, however, may provide direction and assist the environmental professional in isolating the probable allergenic species in a dust sample. It serves well as a screening mechanism. More reliable, easier-to-use results may then be obtained through the monoclonal antigen tests. If, however, the screening fails to disclose any of the common allergens, all is not entirely lost.

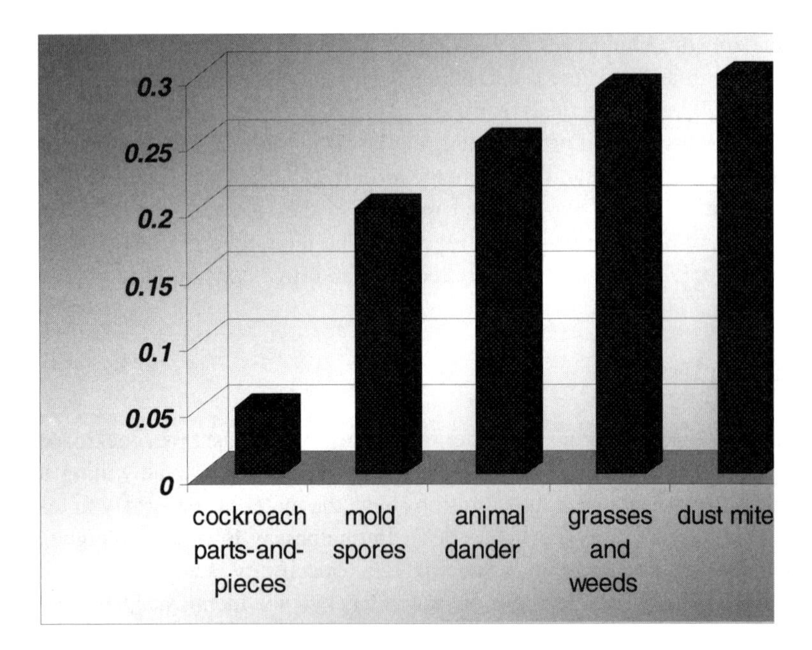

Figure 15.4 Estimated prevalence of allergies in school-age children and skin reactivity tests. Excerpted from *Indoor Allergens*.[16]

An environment may be complicated by rare outbreaks or isolated occurrences. Elevated levels of fleas identified in the carpeting of a high foot traffic, medical reception room cannot readily be assayed. Dust which is known to be the source of allergies cannot readily be identified if the components are other than the more common species. Any of a number of scenarios may develop. In these instances, some laboratories will prepare test material from a given dust sample and test the blood of the allergy sufferers to confirm the allergenic potential of the material. The extracted dust is injected in a strain of mouse. If allergens are present, antibodies are formed. The newly created antibodies are then used to test for dust allergen in an allergy sufferer's blood. Sampling and interpretation require the assistance of a medical doctor.

INTERPRETATION OF RESULTS

A dose-response relationship is recognized with allergens as it is with toxic chemicals. A small percentage of the population will experience the effects at extremely low levels of exposure, levels not even noticeable to a majority of the population. These highly sensitive individuals are thought to comprise less than five percent of the work force. Although this is in line with chemical sensitivity numbers, the more sensitive individuals comprise a group of people who were: (1) predisposed at birth; and (2) exposed to low levels of specific antigens for a long duration. The remaining 95 percent of the population will be impacted as the exposure levels increase.

Ideally, problem area composite samples should be compared with non-problem area composite samples. Composite results may identify suspect allergens, and discrete samples will help to isolate the reservoir. Although most environmental professionals find great comfort in acceptable limits, comparisons with published thresholds should be accomplished with reserve.

Thresholds are in a constant state of flux, and recommended limits may vary from one laboratory to the next. Observations vary by region, and available species differ. Analytical results on the same sample may vary from one laboratory to the next. Laboratory findings and allergenic association tend to be variable, and these findings change within the same laboratory. For this reason, published thresholds should be used as a guide only!

For reference and review, some of the commonly accepted allergy-causing thresholds can be found in Table 15.5. Yet, they are not firm, hard-clad references. These limits are the result of observations made as to measured levels and resultant responses of typically nonatopic individuals. These reference values will not be applicable for those predisposed to allergies. In these atopic individuals (15 to 20 percent of the adult population), their thresholds may be one percent of that which will potentially impact the general population. It should also be noted that immunoassay reference thresholds for plant pollen, mold spores, and chemicals are not

Table 15.5 Allergen Levels in Dust; Capable of Eliciting an Allergic Response in the General Population, Not Predisposed Genetically to Allergies

Allergens	Reference Thresholds (µg/g)
Dust mite allergens (polyclonal) *	15[17]
Der f I	2[17]
Der p I	2[17]
Group I	10[18]; 2[19]
Cockroach allergens **	
Bla g I	2[20]
Cat allergens **	
Fel d I	8[18]; 8[19]
Dog allergens **	
Can f I	—[20]

* *Dermatophagoides farinae* and *Dermatophagoides pteronyssinus*.
** Polyclonal antigens for which there are no published references.

listed. This is due to the direct means available for determining airborne exposure levels for some and/or to the lack of data for reference thresholds.

Another publication sets different range limits, stating that their measurements are the most practical reference values available. They provide the following limits for Group I dust mite allergens:[20]

- Safe levels: less than 2 μg/g
- Levels that may sensitize atopic (genetically pre-disposed) individuals: 2 to 10 μ/g
- Levels that may exacerbate previously sensitized individuals: greater than 10 μ/g

It is thought that the Group I dust mite allergens are associated with fecal material which is relatively small in size as compared with the larger Group II allergens which are associated with the body parts. Thus, the Group I dust mite allergens are those most likely to be disturbed and become airborne. They are the most likely to be associated with allergy symptoms. The Group II body parts may complicate an evaluation where an individual is close to the settled allergenic material (e.g., sleeping on a contaminated pillow or lying on the floor). All conditions must be taken into consideration.[21]

Attempts have been made to report observed thresholds for airborne exposures to rodent allergens. One such attempt is summarized in Table 15.6. As these numbers are variable and not well studied, they should be used with reservations. At best, they may be considered reference guidelines. Dust air sampling should be performed through contributing input from the laboratory that will perform the analysis.

Table 15.6 Allergen Levels in Air Reported to Cause an Allergic Response

Thresholds	Observed Thresholds
Rodent Allergens	
Ag 3	825 μg/m3
Mus m I	59 μg/m^3
Rat n I	no reported observations

Excerpted from *Journal of Allergy and Clinical Immunology.*[22]

OTHER TYPES OF ALLERGENIC SUBSTANCES

Other types of allergenic substances include indoor/outdoor allergens and industry-related allergens. Indoor/outdoor allergens are mold spore allergens.

Mold spore dust sampling is rare. Easier, more commonly accepted approaches are available for analysis (as has been discussed in previous chapters), and there are limited guidelines for interpretation of results. In one report,[23] the researchers suggest that should the fungal spores and bacteria in dust exceed 10,000 colony-forming units per gram of dust (CFU/g) remediation may be indicated. Fungal concentrations on water-damaged materials (e.g., carpeting, gypsum board, or ceiling tiles) are excessive if they exceed 1,000 CFU/g or 1,000 CFU/cm^2. However, the analysis for fungal spores relies on spore viability and is performed by the use of a culture media and Petri dishes. There are no immunoassay methods presently being used for mold spore quantitation in dust, but there are immunoassay methods to determine if an individual is allergic to specific mold spores. History must dictate allergies, and the testing is performed for specific genera.

In certain industries, some of the more common allergenic proteins for which immunoassay methods may be performed have been identified in Table 15.7. Presently, however, processing of the samples may involve considerable expense and/or the aid of a research laboratory. There has been minimal or no dust sample analyses of these materials, and interpretation may be illusive.

Table 15.7 Allergenic Materials Which May Become Airborne

Industry	Type
Food processing	grain, flour, coffee bean, castor bean, egg, garlic, and mushroom dusts
Chemical/industrial	latex proteins, isocyanates, metals, resins, dyes, and drugs

Although each of the above allergens can be analyzed by immunoassay methodologies, many of the chemicals may be sampled using traditional industrial hygiene air and surface sampling methodologies. Once a laboratory has been identified to perform an analysis, the feasibility of performing air sampling for occupational allergens in the food processing industry will be based on one's willingness to collect comparative samples in non-problem areas and compare them to problem areas and/or with symptomology in order to establish an acceptable limit.

Then, also, the environmental professional may encounter a situation whereby dust from a grain elevator or adjacent food processing plant appears to be causing problems for the occupants of a building which is associated only by proximity to the food processing plant. In such cases, source dust samples in the immediate vicinity of a given building may be compared to dust collected indoors. Where symptoms indicate a probable exterior source, some of the occupants may also be tested with the aid of a medical doctor.

Of the chemical/industrial sources, latex protein analysis can be performed by immunoassay of a patient's blood. Drug dispersion sampling is often performed in-house by the pharmaceutical company laboratories, and the other chemicals can be sampled/analyzed as chemicals (not allergens) and compared with published reference limits for toxicity and/or sensitization levels.

Natural rubber latex proteins (produced from the sap of the *Hevea brasiliensis* trees) cause dermatitis to many of those who wear latex gloves, particularly to those in the medical professions where frequent use of the gloves is required. The frequent use and replacement of these gloves in an operating room are reputed to be the potential source of airborne latex proteins in the operating rooms. Of considerable concern, airborne latex proteins in an operating room, where the patient's system is accessible, may result in anaphylactic shock and possible death if the patient is already allergic to latex proteins. Another surprise use and source of airborne latex proteins has been reported in fireproofing material sprayed on structural members of buildings.

In one case, analysis of the fireproofing insulation was performed. The concentrations of latex protein were 1,000 to 2,000 nanograms per gram of material (ng/g). Samples of blood were taken from 36 occupants, and 4 tested positive. The dust tested positive (5 to 26 ng/100 cm^2), and the airborne concentrations ranged from nondetectable to 16 ng/m^3. The only confirmation of the latex allergens as being the cause of skin rashes and other allergic reactions was remediation that led to elimination of the symptoms. Although there were no controls, the latex-containing fireproofing was implicated as the probable source of building complaints.[24]

SUMMARY

Animal allergens that can be identified by the immunoassay methods discussed herein are limited. The investigator should be aware of those that are possible and those that can be analysed. For those that cannot be anaysed by immunoassay, the author suggests forensic dust analysis or guilt by association. For instance, if there are bats residing within a building and the occupants are experiencing allergy symptoms, the investigator may seek to confirm by forensic dust analysis the presence of insect parts-and-pieces. See the cover of this book for a photomicrograph of bat guana.

REFERENCES

1. Burge, Harriet A. *Bioaerosols*. Lewis Publishers, Boca Raton, Florida, 1995. pp. 134-137.
2. Borror, Donald J. and D. DeLong. *Introduction to the Study of Insects*. Holt, Rinehart and Winston, New York, Third Edition, 1971. pp. 634, 636-637.
3. Halsey, John F., Ph.D. and M. Colwell, M.S. Allergy Basics for IAQ Investigations. (Handout) Professional Development Course, American Industrial Hygiene Conference & Exposition, May 20, 1995.

4. Lintner, Thomas J., Ph.D. and K. Brame, B.S. The effects of season, climate, and air-conditioning on the prevalence of *Dermatophagoides* mite allergens in Household Dust. *Journal of Allergy and Clinical Immunology*. April 91(4):862-7 (1993).
5. Burge, Harriet A. *Bioaerosols*. Lewis Publishers, Boca Raton, Florida, 1995. pp. 138-139.
6. Ibid. pp. 151-153.
7. Lintner, Thomas J., Ph.D. Topics on Allergens. [Oral communication] Vespa Laboratories, Spring Mills, PA. October 1995.
8. Burge, Harriet A. *Bioaerosols*. Lewis Publishers, Boca Raton, Florida, 1995. p. 154.
9. Jones, Robert B., et. al. The effect of relative humidity on mouse allergen levels in an environmentally controlled mouse room. *American Industrial Hygiene Association Journal*. 56:398-401 (1995).
10. Burge, Harriet A. *Bioaerosols*. Lewis Publishers, Boca Raton, Florida, 1995. pp. 154-155.
11. Dr. Halsey. IBT Reference Laboratory, Specializing in Environmental Allergen Testing. [Bulletin] IBT Reference Laboratory, Lenexa, Kansas. pp. 2-3.
12. Sylvania. Black Light Radiant Energy. [Engineering Bulletin 0-306]. Sylvania, Danvers, Massachusetts. (1996)
13. Halsey, John F., Ph.D. and M. Colwell, M.S. Allergy Basics for IAQ Investigations. (Handout) Professional Development Course, American Industrial Hygiene Conference & Exposition, May 20, 1995.
14. Lintner, Thomas J., Ph.D., et. al. Sampling dust from human dwellings to estimate the prevalence of *Dermatophagoides* Mite and Cat Allergens. *Aerobiologia*. April 10(1):23-30 (1994).
15. Barbee, R.A., et. al. Longitudinal changes in allergen skin test reactivity in a community populations sample. *Journal of Allergy and Clinical Immunology*. 79:16-24 (1987).
16. Pope, Andrew M., et. al. *Indoor Allergens: Assessing and Controlling Adverse Health Effects*. National Academy Press, Washington, D.C., 1993. p. 52.
17. Chapman, M.D., et. al. Monoclonal immunoassays for major dust mite allergens, *Der p* 1 and *Der f* 1, and quantitative analysis of the allergen content of mite and house dust extracts. *American Academy of Allergy and Immunology*. 80:184-94 (1987).
18. Trudeau, W.L. and E.Fernandez-Caldas. Identifying and Measuring Indoor Biologic Agents. *Journal of Allergy and Clinical Immunology*. August 94(2)2:393-400 (1994).
19. Platts-Mills, T.A.E. Allergen standardization. *Journal of Allergy and Clinical Immunology*. 87:621 (1991).
20. Schou, Carsten, Ph.D., et. al. Assay for the major dog allergen, *Can f* 1: Investigation of house dust samples and commercial dog extracts. *Journal of Allergy and Clinical Immunology*. December 88(6):847-53 (1991).
21. Burge, Harriet A. *Bioaerosols*. Lewis Publishers, Boca Raton, Florida, 1995. p. 135.

22. Twiggs, et. al. Immunochemical measurement of airborne mouse allergens in a laboratory animal facility. *Journal of Allergy and Clinical Immunology.* 69:522 (1982).
23. Trudeau, W.L. and E. Fernandez-Caldas. Identifying and measuring indoor biologic agents. *Journal of Allergy and Clinical Immunology.* 94:393-400 (1994).
24. McCarthy, J.F., K.M. Coghian and D.M. Shore. Latex Allergen Exposures from Fiberproofing Insulation. Presented paper under the heading of Indoor Air Quality at the American Industrial Hygiene Conference. May 22, 1995. p. 6.

GLOSSARY

abscess—A cavity containing pus and surrounded by inflamed tissue.

acid fast—A method of bacterial identification. Acid-fast bacteria retain fuschin stain where other bacterial are rapidly decolorized when treated with a strong mineral acid.

adsorption—Process of collecting a liquid or gas onto the surface of a solid or liquid sorbent.

aerobic—Requiring the presence of oxygen for growth. Some aerobic microbes may form capsules, or spores, when left in an oxygen deficient environment.

aflatoxin—Mycotoxin created by the molds *Aspergillus flavis* and *parasiticus*, a known cancer-causing substance.

algae—Plant-like organisms that practice photosynthesis (requiring light) and, for the most part, live in aquatic environments. They are occasionally implicated with outdoor allergies.

allergic pneumonia—An inflammation of the lungs resulting directly from an allergy, usually to some type of organic dust.

alveolitis—An allergic pulmonary reaction characterized by acute episodes of difficulty breathing, cough, sweating, fever, weakness, and pain the joints and muscles, lasting from 12 to 18 hours.

amoeba—A protozoan which has an undefined, changeable form and moves by pseudopodia (or branching fingers of cellular material).

amplification—In reference to microbes, this means an increase in the numbers (generally due to growth and creation of elevated numbers indoors).

anaerobic—Not requiring the presence of oxygen for growth (e.g., *Clostridium botulism*, which causes canned food poisoning).

angioedema—This is a condition similar to urticaria (or hives), but involves the subcutaneous tissue. It ordinarily does not itch and is a more generalized swelling.

anisotropic—Substances having different refractive indices which depend on the vibration direction of light.

ascospore—A haploid sexual spore created by the fungal class Ascomycetes (e.g., yeasts).

asthma—A disease that is characterized by recurrent episodes of difficult breathing and by wheezing, with periods of nearly complete freedom from symptoms.

atopic—Of or pertaining to a hereditary tendency to develop immediate allergic reactions, such as allergic rhinitis, allergic asthma, and some forms of eczema.

auto immune disease(s)—A person's immune system reacts against its own tissues and organs.

bagassosis—A form of allergic lung disorder caused by exposure to moldy sugar cane fiber.

bake-out—A process whereby the temperature of a building is elevated to force chemical off-gassing of building materials and furnishings. This is generally performed without building occupants, and the air is flushed with outside air after a predesignated time period.

basidiospores—Sexual spores created by the fungal class Basidiomycetes (e.g., common mushroom).

birefringence—The numerical difference in refractive indices for a substance.

booklice (or ***book louse***)—Small, six-legged insects belonging to the order Psocoptera which are implicated in paper-dust allergies.

bronchitis—An inflammation of the mucous membranes of the bronchial tubes, characterized by difficulty breathing.

carcinogenic—A substance capable of causing cancer.

challenge test—A medical procedure, also known as provocative testing, used to identify substances to which a person is sensitive by deliberately exposing the person to diluted amounts of the substance. A positive bronchial challenge is one in which pulmonary function decreases.

commensal—Organisms that live in close association whereby one may benefit from the association without harming the other.

competition—Negative relationships between two populations in which both are adversely affected with respect to their survival and growth.

conidia—Asexually-produced spores which develop at the end of a conidiophore (e.g., *Penicillium*).

conidiophore—The structure from which conidia develop.

conjunctiva—The mucous membrane lining of the inner surfaces of the eye.

conjunctivitis—Inflammation of the conjunctiva. This results in eye redness, a thick discharge, sticky eyelids upon waking in the morning, and inflammation of the eyelids.

croup—An infection of the upper and lower respiratory tract that occurs primarily in infants and young children up to 3 years of age. It is characterized by hoarseness, fever, a distinctive harsh, brassy cough, a high pitched sound when breathing, and varying degrees of respiratory distress.

culturable—Viable (or live) microorganisms which can be grown.

desorption—Process of running adsorbed liquids/gases from sorbent material.

dispersion staining—Staining based upon the differences and similarities of the refractive indices between a solid and liquid.

dispersion—The separation of light into its color components by refractive or diffracted light.

dust mites—Four-legged arachnids which are typically implicated in house dust allergies.

dyspnea—Difficulty breathing.

eczema—A superficial dermatitis which, in the early stages, is associated with itching, redness, fluid accumulation, and weeping wounds. It later becomes crusted, scaly, thickened, with skin eruptions.

emission factor—A single point quantitative measurement of gaseous or particle emissions from a material, as determined by chamber testing.

emission rate—The actual rate of release of vapors/gases from a product over time.

emphysema—A chronic disease of the lungs, in which the alveoli are permanently damaged or destroyed. It is typically characterized by difficult breathing and is associated with heavy, prolonged cigarette smoking.

endocarditis—Lesions of the lining of the heart chambers and heart valves.

endotoxic shock—A body reaction caused by an endotoxin, generally characterized by marked loss of blood pressure and depression of the vital processes.

endotoxin—A toxin produced within bacteria and released upon destruction of the cell in which it was released.

environmental chamber—A nonreactive testing enclosure of known volume with controlled air change rates, temperature, and humidity.

epitope—The portion of an antigen molecule involved in binding to the antibody.

farmer's lung—A form of allergic lung disorder caused by exposure to moldy hay.

flush out—Remove contaminants from a contained air space.

fomites—Inanimate objects (e.g., mineral dust particles).

forensic—Relating to or dealing with the application of scientific knowledge to legal issues.

fungi—Plants that, unlike green plants, have no chlorophyll and must depend on plant or animal material for nourishment.

gastroenteritis—An inflammation of the stomach and intestines. It may at times result from an allergic reaction.

Gram negative—A staining process which is used in diagnostic bacteriology whereby the bacterial wall is stained pink.

Gram positive—A staining process which is used in diagnostic bacteriology whereby the bacterial wall is stained purple, almost exclusively a property of producers of potent exotoxins.

granuloma—A granulated nodule of inflamed tissue.

hapten—A low molecular weight chemical that is too small to be antigenic by itself but can stimulate the immune system when combined with a larger molecule (e.g., protein).

hay fever—Common name for "nasal allergy." Its symptoms include attacks of sneezing, runny, stuffy nose, and itchy, watery eyes, and they occur within a few minutes to a few hours after exposure to inhaled allergens-usually pollen, spores or molds, house dust, or animal dander. The term hay fever is misleading, since these reactions are not usually produced by hay and are not accompanied by fever.

hemorrhagic shock—Physical collapse and prostration associated with sudden and rapid loss of large amounts of blood.

heterologous (cross-reactive) antigen—A different antigen from that which was used to immunize or challenge the immune system, yet it is similar enough to be recognized as the initial foreign substance. Typically, these antigens are polysaccharides, possibly due to their limited chemical complexity and often structurally similar nature to one another. For example, human blood Group B sometimes reacts with antibodies to certain strains of *Escherichia coli*, which is a common bacterial resident in the human colon.

hives (urticaria)—A common skin condition that is probably familiar to everyone. It is a skin rash characterized by areas of localized swelling, usually very itchy and red, and occurring in various parts of the body. It usually lasts only a few hours and involves only the superficial areas of the skin. Urticaria has either an immunologic or a nonimmunologic cause.

homologous antigen—A foreign substance used in the production of antiserum.

house dust—Heterogeneous, firm gray powdery material that accumulates indoors. This category includes mold, pollen, animal dander, food particles, kapok, cotton lint, insects, and bacteria.

hybridized—Bound to a template of DNA sequencing.

hypersensitivity pneumonitis—An allergic disease of the lungs caused by inhaling various allergenic dusts.

hypersensitivity—A condition in which the immune system reacts to antigens that cause tissue damage and disease.

hyphae—A microscopic filament of cells that represents the basic unit of a fungus. They usually do not exist with yeasts.

immune—A condition in which an organism is protected against, or free from, the effects of allergy or infection, either by already having had the disease or by inoculation.

immunoglobulins—One of a family of proteins to which antibodies belong.

in vitro—In an artificial environment.

in vivo—In a live organism.

incidence—Number of new diseases occurring in a given population at a specified time period.

isotropic—Substances showing a single refractive index at a given temperature and wavelength, no matter what the direction of light may be.

itch mites—Achnids with four pairs of legs which cause dermatitis and are sometimes implicated in house dust allergies.

ligase—Protein which acts as a glue to hold bits of DNA molecules together.

lipids—Fats and fatty-like materials that are generally insoluble in water.

lymphocyte—A white blood cell important in immunity. Of the two major types (both types have several subclasses), T lymphocytes are processed in the thymus and are involved in cell-mediated immunity, and B lymphocytes are derived from the bone marrow and are precursors of plasma cells, which produce antibody.

macrophage—A scavenger white blood cell that plays a role in destroying invading bacteria and other foreign material. It also plays a major role in the immune response by processing or handling antigens and as an effector cell in delayed hypersensitivity.

meningitis—An infection or inflammation of the membranes covering the brain and spinal cord. Onset symptoms are characterized by severe headache, stiffness of the neck, irritability, malaise, and restlessness, followed by nausea, vomiting, delirium, and disorientation. This progresses to increased temperature, pulse rate, and respiration. Nerve damage may culminate in deafness, blindness, paralysis, and/or mental retardation.

micron (or micrometer)—One millionth of a meter (10^{-6}).

mold—Microscopic plants, belonging to the Fungi Kingdom, which do not have stems, roots, or leaves and are composed of a vegetative threadlike element (hyphae) and reproductive spores.

monoclonal antibody—Antibody produced by a single clone of cells that binds to only one epitope.

mushroom—Microscopic plants, belonging to the Fungi Kingdom, which are filamentous in nature with large fruiting bodies (referred to as the mushroom cap) which discharge reproductive spores.

mutagenic—Capable of causing a genetic change.

mycelium—A visible mass of tangled filaments of fungal cells, the rooting structure and extension of the more aerial hyphae.

myeotoxins—Toxins produced by molds.

manometer—One one-thousandth of a micron, or one billionth of a meter (e.g., 10^{-9}).

nasal congestion—Blockage of the nasal passages.

necrosis—Localized tissue death.

neutralism—A lack of interaction between two microbial populations.

nonpolar compounds—Compounds which have atoms that do not have a denser electron cloud about one atom, or group of atoms, than around its adjoining atom, or group of atoms (e.g., hexane).

nonviable—Not living, dead, destroyed organisms; not capable of causing disease.

nosocomial—Hospital acquired infections.

paper mites—Fictitious term with possible reference to storage mites or booklice.

parasitism—One population benefits and normally derives its nutritional requirements from the population that is harmed.

patch test—Used to identify substances responsible for contact allergy. The test consists of applying a small amount of a suspected substance to the skin. The area is covered with tape and left for forty-eight hours. It is a small area where the substance was applied swells and turns red, the test result is said to be positive.

Petri plate (or dish)—A glass or plastic dish which has a lid and is used to isolate and grow microbes.

plasmids—Small loops of DNA in bacteria (contain genes for antibiotic resistance).

pleurisy—Inflammation of the pleura (or membranous sacs surrounding the lungs).

polar compounds—Compounds which have atoms that have a denser electron cloud about one atom, or group of atoms, than around its adjoining atom, or group of atoms (e.g., butanol).

pollen—Male fertilizing elements of a plant, which are microscopic in size. Pollen grains are spheroid, ovoid, or ellipsoid in shape and may have a smooth, reticulated, spiculated, or sculptured surface.

predation—One organism, the predator, engulfs and digests another organism, the prey.

predicted air concentrations—A calculated prediction of air concentrations of a given substance or substances, based on component emission rates.

prevalence—Normal frequency of a disease in a given population.

prick method—An allergy test whereby the skin is pricked with a needle at the point where a drop of allergen has been placed, introducing the allergen to the body's immune system.

pyrexia—A condition resulting in fever.

pyrocanic—Sensation of heat.

ragweed—A plant, belonging to the family Compositae, that is the major cause of hay fever in the United States. The ragweed family is large, with approximately 15,000 species.

retractive index—The ratio of the velocity of light in a vacuum to the velocity of light in a given medium. A higher atomic number generally results in a higher refractive index.

restriction enzymes—Proteins that cut apart DNA molecules.

rheumatic fever—An inflammatory disease that usually occurs to young school aged children and may affect the brain, heart, joints, skin, or subcutaneous tissues. Early on, it is characterized by fever, joint pains, nose bleeds, abdominal pain, and vomiting, progressing to chest pain, and, in advanced cases, heart failure.

rheumatoid arthritis—A chronic, destructive, often deforming, collagen disease which results in inflammation of the bursea, joints, ligaments, or muscles. It is characterized by pain, limited movement, and structural degeneration of single or multiple parts of the musculoskelatal system.

rhinitis—A disease of the nasal passages that is characterized by attacks of sneezing, increased nasal secretion, and stuffy nose (caused by swelling of the nasal mucosa).

rust—A fungal disease of agricultural crops so named because of the orange-red color it imparts to infected plants. Belongs to the same fungal class as the common mushroom.

saponification—The hydrolysis of an ester by an alkali, producing a free alcohol and an acid salt.

saprophytic—Lives on and derives its nourishment from dead or decaying organic matter.

scratch test—A small drop of an allergen is applied over an area of the skin where a superficial scratch has been made. This allows the allergen to penetrate the top layer of skin.

semivolatile—Compounds with a boiling range of 240°C to 400°C.

sensitize—To administer, or expose, to an antigen provoking an immune response so that, upon later exposure to that antigen, a more vigorous secondary response will occur. An individual can be immune (e.g., protected against the effects of an infectious agent or antigen) and sensitized to the antigen (e.g., demonstrate a positive tuberculin reaction) at the same time.

septicemia—Systemic infection where pathogens are spread through the bloodstream, infecting various parts of the body, characterized by fever, chills, prostration, pain, headache, nausea, and/or diarrhea.

skin test—A method of testing for allergic antibodies. A test consists of introducing small amounts of the suspected substance, or allergen, into the skin and noting the development of a positive reaction (which consists of a wheal, swelling, or flare in the surrounding area of redness). The results are read fifteen to twenty minutes after application of the allergen.

slime mold—Organisms, belonging to the Fungi Kingdom, which are protozoan-like for part of their life cycle and reproduce by forming stalks that produce spores (or multiple spore-containing sporangia).

smut—A fungal disease of agricultural crops so named because of the sooty black appearance it imparts to infected plants. Belongs to the same fungal class as the common mushroom.

sorbent—A solid or liquid material which collects liquid and/or gaseous substances.

sporangiospores—Asexually produced spores which develop within a sporangium.

sporangium—A structure, or sac, within which spores develop.

spores—Reproductive cells of certain plants and organisms. Inhaled fungal spores are frequently the cause of allergic symptoms such as rhinitis and asthma.

storage mites—Four-legged arachnids which are sometimes implicated in outdoor environmental allergies associated with agricultural environments.

symbiotic—Obligatory relationship between two populations that benefits both populations.

synergistic—Mutually cohabitate with one another.

teleospores—Asexual spores produced by rusts.

thymine—A DNA base which only pairs to the base adenine.

tobacco sensitivity—Many people suffering from rhinitis and asthma experience heightened symptoms when exposed to tobacco smoke.

transparency—Ability to transmit light.

ulceration—Formation of a circular, crater-like lesion of the skin or mucous membranes.

unculturable—Viable and nonviable microorganisms which cannot be grown.

urediospores—Asexual spores produced by smuts.

urticaria (hives)—This is a skin rash characterized by areas of localized swelling, usually very itchy and red, and occurring in various parts of the body. It usually lasts only a few hours and involves only the superficial areas of the skin.

viable—Living organism; capable of causing disease.

viron—An intact virus particle which has the ability to infect.

virus—A submicroscopic organism which consists of genetic material and a coating and requires living organisms in order to reproduce

viscera—The internal organs enclosed within a body cavity, primarily the abdominal organs.

VOC—Volatile organic compound(s).

volatile—Compounds with a boiling range of 0° to 290°C.

yeasts—Species of fungi that grow as single cells.

zeophylic—Dry-loving organism.

Appendix 1

ABBREVIATIONS/ACRONYMS

AAAAI	American Academy of Allergy, Asthma, and Immunology
ACGIH	American Conference of Governmental Industrial Hygienists
AIDS	Auto Immune Deficiency Syndrome
ASHRAE	American Society of Heating, Refrigerating and Air-conditioning Engineers
BA	Blood Agar
CDC	Center for Disease Control
CFU	Colony Forming Units
CFM	Cubic Feet per Minute
CFR	Code of Federal Regulations
DOT	Department of Transportation
DNPH	Dinitrophenylhydrazine
EDXRA	Energy Dispersive X-ray Analyzer
EMA	Electron Microprobe Analyzer
EPA	Environmental Protection Agency
EU	Endotoxin Units
FDA	Food and Drug Administration
FID	Flame Ionization Detector
FTIR	Fourier Transform Infrared Spectrometry
GC	Gas Chromatography
GC/FID	Gas Chromatography/Flame Ionization Detector
GC/ECD	Gas Chromatography/Electron Capture Detector
GC/MS	Gas Chromatography/Mass Spectroscopy
GC/NSD	Gas Chromatography/Nitrogen Selective Detector
HEPA	High Efficiency Particulate Airfilter
HPLC	High Pressure Liquid Chromatography
HRP-SA	Horseradish Peroxidase-Streptavidin
IAQ	Indoor Air Quality
IMA	Ion Microprobe Analyzer
IP	Indoor Pollutant Methods (EPA)
KLARE	Kinetic-Turbidimetric Limulus Assay with Resistant-parallel-Line Estimates (Test)
LAL	Limulus Amebocyte Lysate

MPI	Mass Psychogenic Illness
MVOC	Mold Volatile Organic Compounds
NA	Not Applicable
NAB	National Allergy Bureau (Program under the American Academy of Allergy, Asthma, and Immunology)
NIOSH	National Institute for Occupational Safety and Health
NPD	Nitrogen Phosphorous Detector
OSHA	Occupational Safety and Health Act
OVA	Organic Vapor Analyzer
PID	Photoionization Detector
ppb	parts per billion
ppm	parts per million
ppt	parts per trillion
RBA	Rose Bengal Agar
RT	Room Temperature
SAED	Selected Area Electron Diffraction
SEM	Scanning Electron Microscopy
TCLP	Toxicity Characteristic Leaching Procedure
TEM	Transmission Electron Microscopy
TLV-TWA	Threshold Limit Value—Time-Weighted Average
TO Methods	Toxic Organic Methods
TSA	Tryptic Soy Agar
VAV	Variable Air Volume
VAS	Visual Absorption Spectrometry
VOC	Volatile Organic Compound
WHO	World Health Organization

Appendix 2

UNITS OF MEASUREMENT

VOLUME

1 liter (l) = 1.06 quarts
1 milliliter (ml.) = 10^{-3} liter
1 microliter (µl) = 10^{-6} liter

LENGTH

1 meter (m) = 3.281 feet = 39.37 inches
1 centimeter (cm) = 10^{-2} meter = 0.039 inch
1 millimeter (mm) = 10^{-3} meter
1 micrometer (µm) = 1 micron (µ) = 10^{-6} meter

WEIGHT

1 gram(s) (g) = 0.035 ounce
1 milligram (mg) = 10^{-3} gram
1 microgram (µg) = 10^{-6} gram

TEMPERATURE

1° Fahrenheit (F) = [(1.8) (X°C)] + 32
1° Centigrade (C) = [X°F — 32]/1.8

SAMPLE UNITS

ppm = parts of contaminant per million parts of sample material (e.g., air)
ppb = parts of contaminant per billion parts of sample material

ppt = parts of contaminant per trillion parts of sample material
mg/m^3 = milligrams of contaminant per cubic meter of sample material
$\mu g/m^3$ = micrograms of contaminant per cubic meter of sample material
grains/m^3 = grains of pollen per cubic meter of air sample
CFU/m^3 = colony forming units per cubic meter of air
mg/m^2-hour = milligrams of contaminant per square meter of material in one hour
mg/X-hour = milligrams of contaminant per item, or composite unit, (X) in one hour of emissions testing
mg/hour-m^3 = milligrams of contaminant emitted in one hour within a three cubic meter space
μg/X-hour = micrograms of contaminant per item, or composite unit, (X) in one hour of emissions testing

FLOW RATES

1 cubic centimeter per minute (cc/min.)
1 liter per minute (lpm)

Appendix 3

ALLERGY SYMPTOMS

COMMON ALLERGY SYMPTOMS

Many of the allergy symptoms are commonly known and understood. Yet, some that are not acknowledged by the general public. Improperly interpreted allergy symptom may result in a search for other causes of a given symptom other than allergens. The investigator should be familiar with the symptoms as presented herein.

Allergic Asthma

Asthma generally results when cold air, pungent odors, viral infections, aspirin and related drugs, inert dusts, or allergens cause hyperactivity of the airways with narrowing of the air passages. This is translated into a "tightening sensation of the chest and breathing difficulties associated with coughing and wheezing. Movement of the chest-wall muscles may be felt as a heavy weight, a constriction of the lungs. If complications involve the pneumothorax, rib fracture, or pneumonia, chest tightness becomes pronounced and painful. Breathing could become excruciating and resemble symptoms of pleurisy.

Strenuous exercise may also heighten the severity as exposure levels increase. In severe episodes, the sufferer may also experience wheezing and mental dullness due to reduced oxygen inspired/delivered to the brain.

If a person has been relatively free of asthma until exposure to animals, cut grass, or a virus, and if the exposure is brief, the attack should subside with proper treatment. It may even subside spontaneously.

Allergic Dermatitis

Although acute symptoms of allergic dermatitis are not singularly diagnostic, they characteristically involve itching, redness, swelling, and a scaly rash.

Itching can be intense and may lead to what is known as "weeping" lesions, caused by the serum oozing from the underlying small blood vessels. Typical areas of the body for this to happen are the cheeks, the creases behind the ears, and at the bends of the arms and legs.

Commonly called eczema or atopic dermatitis, the rash can spread enough to become disabling. During the healing stage, the affected skin thickens, becomes dry, and cracks. Some bleeding may also occur during this stage.

A local infection that takes the form of skin boils is serious and should be considered a threat to the comfort of the individual. As their immune system has yet to develop completely, infants and young children may not demonstrate as severe a reaction as an adult. They may simply develop a red, flat, scaly rash without the annoyance experienced by older individuals.

Allergic Rhinitis

Allergic rhinitis is characterized by nasal congestion and sneezing, which are, in turn, associated with irritation of the throat, eyes, and ears.

Eye Pain

Allergic conjunctivitis can induce a painful, burning condition known as "pink eye." Bright light often adds to the discomfort, leading the individual to appear to be afraid of light. Severe watering and squinting may occur.

Hay Fever

When symptoms are seasonal, allergic rhinitis is referred to as "hay fever." Such allergens include pollen and mold spores. Itching of the nasal passages is a common symptom associated with pollen and mold spores. In an attempt to relieve the discomfort, a sufferer frequently performs nose-rubbing rituals and facial contortions. Itching of the roof of the mouth and inner corner of the eyes may further contribute to demonstrations of distress.

Inner Ear Pain and Hearing Loss

Allergic rhinitis is associated with extensive fluid leakage into the middle ear. Fluid buildup causes an aching pain that can temporarily diminish hearing. If untreated, the condition can cause fever and increased pain, and lead ultimately to rupture of the eardrum.

Nasal Congestion

Nasal congestion may result in breathing difficulties due to blockage of the nasal passages. Congestion may also result in a watery, blood-flecked discharge either from the nose and/or back of the throat. Throat irritation may be described as a "tickle" by some allergy sufferers. Drainage and throat irritation provoke coughing and other indirectly related problems.

When nasal blockage occurs, allergy sufferers tend to breath through their mouth and frequently to lose sleep. The problem thus culminates in fatigue and irritability. Although these symptoms are associated with allergies, it is important that the environmental professional be aware that they may be the result of other problems that may require a physician.

Not all nasal discharges are caused by allergies. Simple colds or virus infections may simulate allergic disease. Allergic disease is not normally accompanied by the presence of fever, general aching, and a yellow or green nasal discharge. When exposed to unusually high levels of an allergen, an allergy sufferer may develop a systemic reaction where general aching is the primary symptom. Yellow or green nasal/throat discharges are generally associated with viral or bacterial infections, and an infection lasts about two weeks. An allergic condition comes and goes with environmental changes.

Another cause of congestion that should not be confused with allergic rhinitis is that which is caused by a tumor obstruction. Nasal blockage may result from polyps—nonmal malignant growths of sinus and nasal tissue filled with fluid, which may or may not be the result of allergy-caused post nasal drip. If the blockage persists unabated for several days, both loss of smell and nasal/sinus infections may result.

Sinus Headache

Sinus obstruction results in area related headaches. The sinuses have outlets or air spaces through the nasal passages. These air spaces are located near the nose, under the cheeks, and above the eyes. When the spaces become obstructed, buildup of air or fluid leads to increased pressure, resulting in pain. This pain may range from a dull discomfort to a sharp, steady ache in the areas involved (e.g., above the eyes). This is referred to as a sinus headache, or "sinusitis."

OTHER ALLERGY-RELATED DISEASES

Some rare forms of allergy-related illness occur in less than one percent of the United States population. Awareness concerning some of their symptoms may be relevant to an overall investigation of environmental impact of airborne allergens. These are discussed herein.

Allergic Bronchopulmonary Aspergillosis

Allergic bronchopulmonary aspergillosis is, as the name implies, an allergic disease that involves exposures to the fungus *Aspergillus*. *Aspergillus fumigatus* is generally the species implicated. This is a disease involving invasion of the mucous lining of the air passages within the lungs. Affected individuals may experience a persistent cough associated with sputum, asthma-like symptoms, fever, and chest pain. Viable spores are the initiator, and, once invasion has occurred, the disease is progressive until treatment is administered.

Contact Dermatitis

Contact dermatitis involves an itchy, red rash which is generally confined to the site of exposure. Scratching may lead to blistering. Although the areas impacted are commonly exposed skin surfaces, exposures may result from transference of allergen from one place to another (e.g., scratching unexposed skin surfaces with contaminated fingernails). The most common exposures are to oils from poison ivy and poison sumac, which if burned may become airborne, or are spread by contact with weeping sores. Removal from and avoidance of the allergen will, in most cases, provide relief within two weeks if the affected areas are left alone.

Hives

Hives, or "urticaria," is generally associated with food or drugs. On rare occasions is may be associated with animal allergens, skin contact with the mold *Penicillium*, and *Hevea brasiliensis* latex. Raised bumps may appear anywhere on the body. These may range in size from that of a small pea to large sections of the body and are often accompanied by red haloes and itching. Each lesion may last only a few hours, but new ones can appear at frequent intervals, and itching may become so intense that it is difficult to perform simple tasks or sleep.

Hypersensitivity Pneumonitis

Hypersensitivity pneumonitis, or "allergic alveolitis," generally involves an occupationally-related inflammation of the alveoli and bronchioles of the lungs. The acute disease, pursuant to extremely high exposures, is characterized by breathlessness, chills, and fever. Fever may be as high as 104°F, and, within four to six hours, symptoms progress into muscle aches and moodiness. Chest X-rays may show diffuse nodular shadows which are predominately found in the lower portions of the lungs.

Subacute symptoms occur upon repeated exposures and appear as chronic bronchitis. This phase is characterized by repeated exposures and is characterized by recurring cough, breathlessness, weight loss, and malaise.

Chronic exposures to low levels of antigenic material may be demonstrated by less obvious symptoms. The allergy sufferer may only experience breathlessness and weight loss, but the chest X-rays will reveal progressive scarring. Pulmonary function tests will show declining functional lung volume. Diagnostic confirmation of the various stages of the disease generally requires the aid of a physician. Common occurrence of the disease is typically limited to bird handlers and to farmers in the southeast. Disease is generally associated with a known source and is region dependent.

Appendix 4

CLASSIFICATIONS OF VOLATILE ORGANIC COMPOUNDS

ALCOHOLS

Due to its association with cosmetics, the more commonly found alcohol in indoor air quality is isopropanol. Alcohols are most frequently used as solvents, perfumes, components of pharmaceuticals (e.g., cough medicine), cleaning agents (e.g., packaged as hand wipes), and fuel additives. The aliphatic alcohols have a distinct "alcohol" odor, and the aromatic alcohols have more of a medicinal or creosote odor.

Low level exposures to aliphatic alcohols may result in irritation of the eyes and upper respiratory tract, occasionally headache and dizziness. The lowest 8-hour exposure for volatile alcohols is 1 ppm (ACGIH limit for propargyl alcohol).

Low level exposures to alicyclic and aromatic alcohols may result in irritation of the eyes and upper respiratory tract, anorexia, weight loss, weakness, and muscle aches and pain. The lowest 8-hour exposure for volatile alcohols is 5 ppm (ACGIH limit for phenol). All alcohols are readily absorbed through the skin barrier and enhance absorption of other chemicals.

ALDEHYDES

The aldehydes are commonly encountered in indoor air quality. They are most frequently used as cleaning agents, biocides, and constituents of resins. Odors have been described as sharp, pungent.

The aldehydes follow the "First Member Rule" in that formaldehyde is the most toxic of the group. Although not as common as many people would suspect, formaldehyde has been known to cause skin sensitization which may demonstrate itself in a fashion like poison ivy, and formaldehyde is a suspect human carcinogen.

Low level exposures to aldehydes may result in irritation and/or a burning sensation of the eyes, upper respiratory tract, and skin. The lowest exposure limit for aldehydes is a ceiling of 0.3 ppm (ACGIH limit for formaldehyde). The ceiling level should not be exceeded for any period of time. For ceiling limits, monitoring is generally performed for 15 minutes or with a direct reading instrument which provided real time data.

ALIPHATIC AND ALICYCLIC HYDROCARBONS

Many of the VOCs found in indoor air quality are aliphatic (e.g., methane) and alicyclic (e.g., cyclohexane) hydrocarbons. Methane, an aliphatic, is evolved during the animal and plant decay, and it is encountered in the atmosphere typically at levels of 1.8 ppm. Many of the aliphatics and alicyclics are components of gasoline, and they are used for heating, welding, illumination gases, refrigeration, and solvents. The aliphatics and alicyclics do not have an odor.

With a low toxicity, the principal effect of aliphatic and alicyclic VOCs is narcosis at very high exposure levels, in excess of what would normally be found in indoor air quality. Symptoms may include lightheadedness, headache, and irritation of the eyes and nose. The lowest 8-hour exposure for volatile halogenated hydrocarbons is 50 ppm (ACGIH limit for n-hexane).

AROMATIC HYDROCARBONS

Most of the more commonly encountered organics in indoor air quality are volatile aromatic hydrocarbons. Aromatics are most frequently used in solvents, and they are a component of gasoline. They have a pleasant, "aromatic" odor.

The aromatic hydrocarbons follow what is referred to as the "First Member Rule." This rule states that the toxicity of the first member of a homologous series is likely to differ qualitatively from the toxicity of the other members of the same series. For instance, benzene, the first member in the aromatic hydrocarbon series, is the more toxic of the aromatics, causes damage to the blood cell-forming system, and is a suspect human carcinogen. The other aromatics are neither as toxic, nor known to cause the same extreme health effects as benzene.

The health effects associated with aromatic VOCs at low levels anticipated in non-industrial environments are narcosis and moderate irritation. Symptoms may include fatigue, weakness, dizziness, headache, and irritation of the eyes, nose and throat. The lowest 8-hour exposure for volatile halogenated hydrocarbons is 0.5 ppm (ACGIH limit for benzene).

ESTERS

Esters are occasionally encountered in indoor air environments. They are most frequently used as solvents, and the most commonly anticipated ester in office environments is ethyl acetate in highlighter pens.

Esters have sweet, variable odors (e.g., bananas or wintergreen).Low level exposures to esters may result in irritation of the eyes and upper respiratory tract and narcosis. Symptoms may include headache and throat irritation as well. The lowest exposure 8-hour exposure limit for ketones is a 5 ppm (ACGIH limit for diisobutyl ketone).

ETHERS

The ethers are rarely encountered in indoor air environments. They are a narcotic, frequently used for anesthesia, and have been used in solvents, making gun powder, refrigerant, and aerosol propellant. Odors have been described as Low level exposures to glycol ethers may result in irritation of the eyes and upper respiratory tract. The lowest 8-hour exposure for volatile alcohols is 400 ppm (ACGIH limit for ethyl ether).

GLYCOL ETHERS (I.E., CELLOSOLVES)

The glycol ethers are occasionally encountered in indoor air environments. They are most frequently used as solvents (referred to by the trade name of Cellosolves) in dry cleaning, nail polishes, varnishes, and enamals. Odors have been described as sweat, ester-like, and musty. Low level exposures to glycol ethers may result in irritation of the eyes and upper respiratory tract. The lowest 8-hour exposure for volatile glycol ethers is 5 ppm (ACGIH limit for 2-ethoxyethanol).

HALOGENATED HYDROCARBONS

Other than the ubiquitous Freon, the volatile halogenated hydrocarbons less frequently encountered in indoor air environments. They are most frequently used as solvents. Halogenated hydrocarbons have a faint sweet odor.The halogenated hydrocarbons follow the First Member Rule in that halogenated methanes (e.g., methyl bromide and chloroform), ethanes (e.g., vinyl bromide), and benzenes (e.g., benzyl chloride) are the more toxic of the group.They are narcotic, can cause liver and kidney damage, depression of the bone marrow activity, and are suspect human carcinogens. Low level exposures to all other halogenated hydrocarbons may result in headache, dizziness, and nausea. The lowest 8-hour exposure for volatile halogenated hydrocarbons is 0.5 ppm (ACGIH limit for vinyl bromide).

KETONES

Ketones are occasionally encountered in indoor air environments. They are most frequently used as solvents. They are also used in antifreeze solutions and hydraulic fluids. The most commonly found ketones in indoor air are acetone and methyl ethyl ketone. Ketones have a faint pleasant odor.Low level exposures to ketones may result in irritation of the eyes and upper respiratory tract and narcosis. Symptoms may include headache and throat irritation as well. The lowest exposure 8-hour exposure limit for ketones is 25 ppm (ACGIH limit for diisobutyl ketone). Ketones are highly flammable.

GENERAL INDEX

Symbols

A

D

E

F

G

H

I

J

K

L

M

P

Q

R

S

T

U

SYMPTOMS INDEX

D

E

F

G

H

I

S

T

U

V

W